Gender, Identity, and Imperialism

COMPARATIVE FEMINIST STUDIES SERIES
Chandra Talpade Mohanty, Series Editor

PUBLISHED BY PALGRAVE MACMILLAN:

Gender, Identity, and Imperialism

Women Development Workers in Pakistan

Nancy Cook

First published in 2007 by
PALGRAVE MACMILLAN™
175 Fifth Avenue, New York, N.Y. 10010 and
Houndmills, Basingstoke, Hampshire, England RG21 6XS
Companies and representatives throughout the world.

PALGRAVE MACMILLAN is the global academic imprint of the Palgrave Macmillan division of St. Martin's Press, LLC and of Palgrave Macmillan Ltd. Macmillan® is a registered trademark in the United States, United Kingdom and other countries. Palgrave is a registered trademark in the European Union and other countries.

ISBN-13: 978–1–4039–7991–9
ISBN-10: 1–4039–7991–X

Library of Congress Cataloging-in-Publication Data is available from the Library of Congress.

A catalogue record for this book is available from the British Library.

Design by Newgen Imaging Systems (P) Ltd., Chennai, India.

First edition: December 2007

10 9 8 7 6 5 4 3 2 1

Printed in the United States of America.

Contents

Series Editor's Foreword

The Comparative Feminist Studies (CFS) series foregrounds writing, organizing, and reflection on feminist trajectories across the historical and cultural borders of nation-states. It takes up fundamental analytic and political issues involved in the cross-cultural production of knowledge about women and feminism, examining the politics of scholarship and knowledge in relation to feminist organizing and social movements. Drawing on feminist thinking in a number of fields, the CFS series targets innovative, comparative feminist scholarship, pedagogical and curricular strategies, and community organizing and political education. It explores and engenders a comparative feminist praxis that addresses some of the most urgent questions facing progressive critical thinkers and activists today. Nancy Cook's *Gender, Identity, and Imperialism* is an excellent example of such comparative feminist praxis. It is located at the intersection of feminist sociology and anthropology, postcolonial studies, critical development studies, discourses of subjectivity and agency, and gendered studies of imperialism, where it carefully and provocatively engages some of the central interconnected issues in these fields.

Over the past many decades, feminists across the globe have had various successes—however, we inherit a number of the challenges faced by our mothers and grandmothers. Moreover, there are new challenges to face as we attempt to make sense of a world indelibly marked by the failure of postcolonial capitalist and communist nation-states to provide for the social, economic, spiritual, and psychic needs of the majority of the world's population. In the year 2007, globalization has come to represent the interests of corporations and the free market rather than self-determination and freedom from political, cultural, and economic domination for all the world's peoples. The project of U.S. Empire building, alongside the dominance of corporate capitalism kills, disenfranchises, and impoverishes women everywhere. Militarization, environmental degradation, heterosexist State practices, religious fundamentalisms, and the exploitation of women's labor by capital all pose profound challenges for feminists at this time. Recovering and remembering insurgent histories has never been so important at a time marked by social amnesia, global consumer culture, and the world-wide mobilization of fascist notions of 'national security.'

These are some of the challenges the CFS series addresses. The series takes as its fundamental premise the need for feminist engagement with global as well as local ideological, historical, economic, and political processes, and the urgency of transnational dialogue in building an ethical culture capable of withstanding and transforming the commodified and exploitative practices of global culture and economics. Individual volumes in the CFS series provide systemic and challenging interventions into the (still) largely Euro-Western feminist studies knowledge base, while simultaneously highlighting the work that can and needs to be done to envision and enact cross-cultural, multiracial feminist solidarity.

Nancy Cook's *Gender, Identity, and Imperialism* extends, complicates, and pushes the range of scholarship in the CFS series to new levels. This ethnographic study of Western development workers' construction and negotiation of gendered, racialized, and imperial subjectivities in Gilgit, Pakistan is an elegant example of what the author calls a "feminist sociology of imperialism." Working against the two dominant narratives of Western women as guiding force or benign presence in masculinist colonial and imperial spaces in South Asia, Cook develops a nuanced account of white women's subjectivities in the context of development work, and the inheritance of colonial discourses of self/other, superiority/inferiority, liberated/oppressed. Thus, the local/global power-filled spaces of development workers' daily lives in Gilgit are explored, analyzed, and theorized with a light, respectful touch that keeps the complexity and humanity of these women always visible. By exploring the private and public spaces of everyday work, leisure activities, and domestic life, Cook's narrative foregrounds the continuities and discontinuities of gendered imperial identities. Cook writes fluently in the languages of social ethnography, cultural studies, geography, philosophy, and postcolonial feminist studies. Besides its comparative, interdisciplinary framing, *Gender, Identity, and Imperialism* is a significant contribution to feminist studies of imperialism, of whiteness, and of South Asia.

This project can open up a number of important theoretical and methodological questions regarding the gendering of imperial subjectivity, the politics of knowledge in the context of development work in the Global South (and its sometimes inadvertent continuities with colonial discourses), and interdisciplinary theorizations of agency in the way subjectivities are constituted through spatial boundary-making and negotiation across sex, race, class, nation, and religion.

This is the kind of scholarship that can create the ground for cross-racial dialogue among and between feminist scholars and activists in regional as well as global contexts. The book will be of interest to a wide range of feminist scholars, ethnographers, cultural critics, and development theorists. It embodies the comparative praxis and vision of transnational knowledge production that is a hallmark of the CFS series.

CHANDRA TALPADE MOHANTY
Series Editor, Ithaca,
New York

Acknowledgments

I have accumulated debts to many people in the course of researching and writing this book, which made its first appearance as my PhD dissertation. First, deepest thanks go to my research participants for their time, friendship, honesty, and enthusiasm for this project. Although my analysis is unsettling, I hope I have succeeded in evoking your lives in Gilgit with sensitivity and careful consideration. Second, I am grateful for the support of the faculty and graduate students in the Sociology and Philosophy departments at York University. I especially want to thank Ratiba Hadj-Moussa, Lorraine Code, and Fuyuki Kurasawa for their scholarly insights, intellectual stimulation, and frank criticism. Jackie Smith and Lorraine Pannett provided further inspiring conversation and thought-provoking questions as they put aside their own work to make substantive comments on later drafts. Third, I could not have completed this research without the material support of the Social Sciences and Humanities Research Council of Canada in the form of a Doctoral Fellowship and York University through two Research Cost Fund grants. Thanks also to Loris Gasparotto, the cartographer at Brock University. Fourth, several people have transgressed all boundaries of care and friendship to create for me a supportive environment in which to undertake this research. Foremost among them in Pakistan are Mehboob Aziz, Abdul Qayuum, Nusrat Nasab, Abdul Bari, and the Yasoob family. Jean Cook, Wally and Dorothea Butz, Kathryn Besio, Ellen Rodger, and Jim Chernishenko sustained me back home. And finally, I owe ineffable thanks to Nina, who offered welcome distractions, and David, who nurtured me in Gilgit and discussed the research at every stage with patience, insight, and sensitivity. This book is dedicated to them.

This book elaborates analyses I put forward in earlier work, which is included here with permission:

2005. What to Wear, What to Wear?: Western Women and Imperialism in Gilgit, Pakistan. *Qualitative Sociology* 28(4): 351–369.

2006. Bazaar Stories of Gender, Sexuality, and Imperial Space in Gilgit, Northern Pakistan. *ACME: An International E-Journal for Critical Geographies* 5(2): 230–257.

2006. Dealing with Danger: Spatial and Mechanical Manipulations in Gilgit, Pakistan. *Gender, Technology, and Development* 10: 191–210.

2007. Gendering Globalization: Imperial Domesticity and Identity in Northern Pakistan. *Institute on Globalization and the Human Condition Working Paper Series.* http://globalization.mcmaster.ca/wps.htm.

Gilgit District, Northern Areas

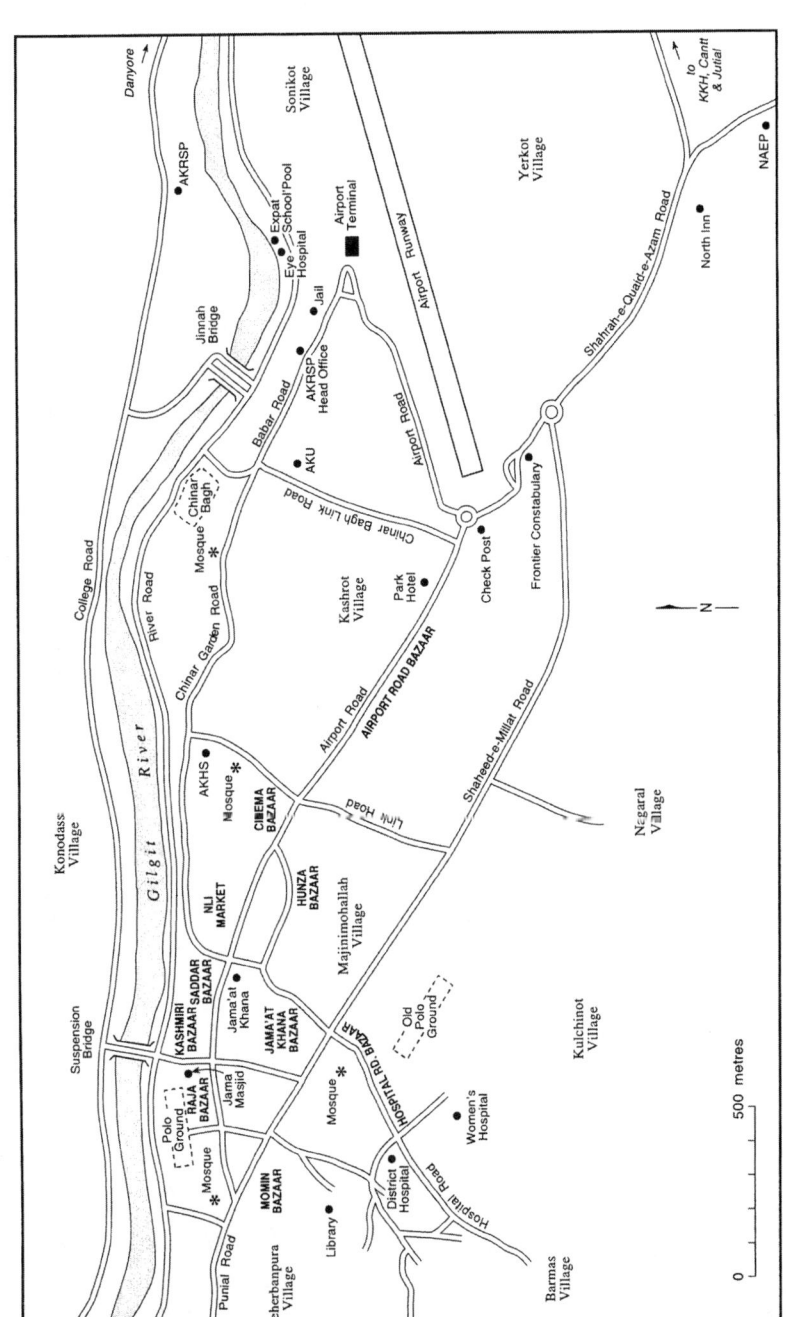

Gilgit City

Introduction: Points of Arrival and Departure

I've learned here that even horror experiences always have a conclusion, and because development volunteers are development volunteers, those horrors are all set in the larger context of a really rich cultural experience.

—Janet[1]

I guess I grieve a little bit or feel sad that I'm not who I used to be. I never will be, because I've lived here. I'm never going to be like I was seven years ago. You know, you start getting mixed up, wondering who you really are. Well, I'm also getting older, so you change because of that, but we don't know anymore what part is getting older and what part is living in Gilgit, becoming more international and world-minded.

—Abbie

Judging by facial expressions, nobody waiting for the bus was looking forward to the long overnight journey from Gilgit to Islamabad in the steamy heat of late August. The bus would be crowded, the ride hard and bumpy, and the young driver invariably sleepy from overwork. Bus rules dictate that the two passengers who sit in the front seat with the driver cannot sleep during the 16 hour journey; it is their responsibility, together with boisterous Bollywood film music, to keep him stimulated and awake through the night. Many people vie for the job, rather than rest fitfully, worrying about him careening off the road that barely hugs the steep mountain slopes.

This bus journey is always exhausting, but anticipation eases the trip for me. Every time I leave Gilgit for home I look forward to, not only a few days of good eating and air conditioned comfort in Islamabad before the flight to Toronto, but also the return to family, friends, modern conveniences, my own schedule, and a way of life that is familiar, reassuring, comfortable. Privilege infuses even this aspect of my traveling; after my holiday or research season is over, I can leave the social conditions my Gilgiti friends complain about—the widespread poverty, the poor infrastructure, the paucity of good schools and health care, the shortage of tourists since Pakistan's nuclear testings and the World Trade Center attacks—while they struggle on with their lives. As I prepare to leave, I routinely consider the ways in which

Westerners could come to better appreciate how particular power structures configure their arrivals to, sojourns in, and departures from Pakistan.

Several feminist scholars across disciplinary boundaries have addressed this question as it relates to Western women who traveled to colonized territories. Literary theorists, geographers, and historians have explored the ambivalent, contingent, multiple, and shifting discourses of power that organize a diverse range of Western women's published colonial-era literature, including the journals, novels, advice manuals, and travel narratives they wrote to describe their residence aboard (e.g., Blunt 1994b, 1999b; Clancy-Smith 1998; George 1994; Ghose 1998; Grewal 1996; Jayawardena 1995; Jayaweera 1990; McEwan 1994; Mills 1996; Nair 1990; Paxton 1992; Spivak 1985; Ware 1992). Although this focus on metropolitan women who wrote from the colonies has been innovative and effective in addressing the power relationships that configured their lives abroad, little effort has, so far, been devoted to understanding how Western women are implicated in contemporary transcultural power relations and what relationship those processes have to the power relations exercised in the colonial era. This book, which is inspired by a concern to think through my own complicity in these processes as a Western tourist and researcher in Pakistan, provides such an analysis by examining the efforts made by Western women development workers living in present-day Gilgit, northern Pakistan to forge comfortable identities and everyday lives, which are constituted through local and global relations of power as manifest in discursive practices of gender, race, class, sexuality, and imperialism.

Joining a host of Westerners who have traveled to Gilgit over the centuries as traders, explorers, conquerors, researchers, adventurers, and development workers, I arrived as a tourist there for the first time in 1986. Although my initial bus trip was unnerving, it was relatively comfortable compared to the treks made by my nineteenth-century counterparts who used well-established mountain paths often too narrow for mules. This improved comfort is facilitated by the Karakorum Highway (KKH), which opened in 1978. It connects the federally administered Northern Areas region of Pakistan, of which Gilgit is a part, to southern Pakistan and Xinjiang province in China via a paved road. The KKH was built cooperatively by the Chinese and Pakistani armies over 13 years at great human cost, as it twists through unstable and high mountainous terrain, following the Indus, Gilgit, and Hunza rivers through the Western Himalayan and Karakorum mountain ranges to China along one of the ancient silk routes of Central Asia. Since the first century AD this region has been an important hub of trade and political activity. During the colonial era it was vital to British military efforts to compete against Russia for political influence in Central and South Asia in what is popularly called the Great Game.

Although this region once belonged to the British Raj as part of Kashmir, it was sold to the Raja of Jammu in 1846. When India was partitioned in 1947, the Northern Areas were designated part of Hindu-ruled Indian Kashmir. The Maharaja of Kashmir, under pressure from Nehru, opted to join India, not Pakistan, even though Kashmiris are predominantly Muslim. As proud Gilgiti historians relate, the Muslim population staged a coup against the Kashmiri government over this decision, and the area was consequently declared the Independent Republic of Gilgit. The Northern Areas and Azad Kashmir acceded to Pakistan in 1949, but the rest of Kashmir remains disputed territory. A United Nations ceasefire has been in effect since 1949, but fighting between India and Pakistan, including three major military clashes in 1965, 1971, and 1999, is a regular occurrence along the disputed border.

Gilgit, which is the largest town in the Northern Areas, lies in a steep valley at 1400 meters above sea level at the confluence of the Hindukush, Karakorum, and Himalaya mountain ranges and the Hunza and Gilgit Rivers. It is a refreshing sight at the end of the 700 kilometer bus ride from Islamabad along the KKH. Despite the dusty landscape and desert mountain environment, Gilgit has been converted into a lush oasis with patches of green gardens, terraced fields, and orchards of apricot, cherry, and apple trees by local landowners who work vegetative miracles as they irrigate land with water from melting snows from higher altitudes.

Gilgit town was almost immediately affected by the increased, if intermittent, economic trade and human mobility that the KKH allows, social flows that initiated state attempts to integrate a largely subsistence agricultural society into the national economy (Allan 1985; Dittman 1997; Kamal et al. 1997; Kreutzmann 1991). The city became the administrative, military, and economic center of the Northern Areas, attracting Punjabi, Pathan, Tajik and Uyghur traders, various international development organizations, national and international tourists, and village people from all over the north looking for nonagrarian jobs. As a result, the population of Gilgit has grown dramatically. The 1981 census estimated the population at 29,102 (Streefland et al. 1995). In 2002, town administrators report, the number had risen to approximately 60,000.

The KKH has accelerated the pace of development and growth in Gilgit, but it has also put stress on some social and cultural aspects of Gilgiti life. First, Gilgit receives less than 130mm of rainfall every year (Iturrizaga 1997). Water must be channeled from the Kargah Valley, seven miles to the north of Gilgit, for household use, irrigation, and hydroelectric power (Dani 1991). The growing population in the area has created household and agricultural water needs, as well as electricity and sewage demands, that the annual snow melts cannot sustain.

Second, Gilgit has, over the past two decades, become a site of religious tensions. Collective attempts by Sunni, Shia, and Ismaili[2] migrants to live and work in religiously and ethno-linguistically segregated neighborhoods and bazaars are increasingly frustrated as the population and markets grow, sprawling into one another. In May 1988, for instance, Sunnis attacked their Shia neighbors with guns, ostensibly because the Shias blasphemously finished the Ramadan month of fasting one day ahead of Sunnis. Hundreds of people, from both sects, were killed. Again in November 1989, Sunnis left several Shias dead after a similar religious disagreement, and for several months in early 2005 Gilgiti citizens lived under curfew owing to renewed sectarian violence. Ismaili people, who are generally much less concerned than their neighbors about fasting, public praying, clothing and grooming regimes, and women's spatial seclusion, fear attacks by both Sunnis and Shias, many of whom do not consider them to be 'real' Muslims. Many Ismaili women behave far more conservatively in Gilgit than they do in their home villages to prevent sectarian assaults against their families and communities.

Although Pakistan International Airlines subsidizes daily flights to Gilgit in small Fokker Friendship planes that navigate around nearby Nanga Parbat (8,125m) without the aid of radar, due to cloudy weather, only once, in my six visits, have I managed to arrive in Gilgit by air. The KKH is the most reliable route, even when it is frequently blocked by mudslides and rock avalanches that army personnel clear with bulldozers. Any number of patient travelers share tea and biscuits at these road blocks; Gilgiti students and businessmen, down-country Pakistani tourists headed to the cooler climate of the hills, and international development workers wait together for the debris to be cleared.

The newly constructed KKH advantages not only these travelers but also international development agencies. The physical access to mountain communities it provides has allowed the Aga Khan Rural Support Program (AKRSP)—a branch of the Aga Khan Development

Network (AKDN)[3]—to consolidate and intensify development activity in the villages of Gilgit District, a project it began in December 1982. Next to the army and the civil administration, AKRSP is the largest employer in the region, and its Northern Areas head office is in Gilgit. Along with numerous other development projects,[4] it employs many local people and down-country Pakistanis, as well as Western student interns, volunteers, consultants, and General Managers. This wide-ranging development activity explains why there were approximately 75 Westerners living in and around Gilgit in 1999 and 2000.

As I arrived for my second research field season in the spring of 2000, several weeks had passed since these development workers had hosted a party. They were absorbed in running teacher training courses in the outlying districts of Gilgit, conducting workshops, equipping new health clinics, performing eye surgeries, arranging cultural festivals, setting up education consultancies, helping women deliver babies, and conducting evaluation research on recently implemented agricultural initiatives. No one had the energy or the time to organize a gathering. But when too many birthdays had passed uncelebrated and everyone was eager for a bit of fun, Allan threw a Western-style house party, complete with pasta dinner, imported alcohol, and dancing.

A handful of Gilgiti men attended alongside approximately 30 foreigners. I was acquainted with all of the women, as they were my research participants. Over the past two summers we had become well acquainted through formal interviews, casual conversations, and socializing opportunities such as this. From what I had heard and witnessed in the past, I guessed they would be ready to let loose after all these weeks, to celebrate each other's company and culture within one of Gilgit's Western microcosms. Despite the heat and inquisitive local onlookers, when the music started, the women rushed onto the dance floor in a crush. If their energy had waned toward the end of the evening, it was revived by the first few bars of the Bee Gees' disco hit *Stayin' Alive*. Lyn, who was singing at the top her of lungs while she danced with me, shouted over the music, "This song is interesting for your research. It should be our theme song really. That's just what we're trying to do. Western women are just trying to stay alive in this place."

Lyn was alluding to the difficulties most white Western women[5] experience in constructing comfortable lives and identities in this socially unfamiliar Pakistani frontier town. Settling into a life abroad is stressful, especially when unaccustomed languages and social interaction rules, as well as a different religion, keep you ill at ease, making you feel uprooted without coherent guidance about how to negotiate the social setting in which you expect to live and work for several years. During my previous trips to Gilgit, all the Western women I spoke to, like me, were anxious about making a life there. They were concerned about how to dress, where to live, where to find medical treatment, what types of transportation to use, and where to send their children to school. They fashioned long lists of 'dos' and 'don'ts'— according to their understandings of Gilgiti people, customs, and spaces—to shape comfortable daily routines and satisfactory identities for themselves. Expatriate parties serve as a way of coping with these stresses and uncertainties; they rejuvenate women, keep them feeling alive, allow them to preserve a familiar identity as modern, white, middle-class Western women who dance to disco music, drink alcohol, wear jeans and t-shirts, live in well-appointed homes, and eat pasta with leavened bread. I was interacting with this group of women in order to understand their efforts to 'stay alive' in Gilgit. More specifically, over time, and in concert with reflections on my own conduct, I became interested in the ways in which they constitute their subjectivities in this transcultural setting through particular discourses of power that organize their self-imaginings and sociospatial practices of inclusion and exclusion, including expatriate house parties.[6] My curiosity extended to understanding how, through these processes of subjectivity formation, Western women perpetuate, legitimate, resist, and transform relations of domination as they imagine themselves in relation to the people among whom they live, construct communities and homes, and build careers and relationships in Gilgit.

As foreign development workers have come to Gilgit to initiate Western-derived sociocultural reforms among a once-colonized group of people, an analogy between them and Raj-era memsahibs is plausible. However, I am reluctant to refer to my research participants as modern day 'lady masters.' Although the dominant trope represents colonial Western women as insignificant or marginal to the masculine enterprise of Empire except as nurturers of their male counterparts (Levine 2004a; Procida 2002; Sharpe 1993), many expatriated British sahibs have depicted these women as domestic ogres, who, through their tyrannical rule of the colonial household, are responsible for undermining imperial race relations and thus orchestrating the failure of Empire (Allen 1976; Barr 1976; Callaway 1987; MacMillan 1988; Nadis 1957; Spear 1963; Strobel 1991). Rather than perpetuate this latter myth that Western women constitute the heartbeat of colonialism, or the former and equally problematic notion that their 'civilizing' presence in colonial and postcolonial[7] contexts is benign, I aim to explore the ambivalent, contingent, multiple, and shifting discourses of power that organize a diverse range of Western women's subjectivities and practices in contemporary Gilgit.[8]

Several postcolonial[9] scholars have recently called for such an investigation. Robert Young (1990), for example, has argued that much work, led by the insights of Edward Said (1978), has employed colonial discourse analysis to reveal enactments of power in the colonial context. The next step should involve "an analysis extended to the discursive formations, representations, and practices of power in contemporary social contexts, together with their relations to the colonial past and to nineteenth-century forms of knowledge, showing how they sustain and intervene in contemporary practices which legitimize [various forms of domination]" (Young 1990, 175; see also Said 1989). Sarah Mills (1996) urges social scientists with a focus on the material aspects of postcolonial relations to go beyond postcolonial literary and cultural theory that restricts its interests mainly to written or artistic texts produced in the colonial era. Finally, Shohat and Stam (1994, 2), in their analysis of Eurocentrism, argue that it is essential to make visible the forms of "vestigial thinking which permeate and structure contemporary practices and representations even after the formal end of colonialism." Drawing on the methodological tools of sociology and anthropology, I will demonstrate that within the setting of postcolonial Gilgit, circulating discursive frameworks are the contingent legacy of the colonial period and the norms, values, and sociospatial boundaries that were developed and contested at that time.

Colonialism and imperialism are two discursive regimes crucial to my analysis. The distinction between them often tends to collapse because they are such closely overlapping and mutually constitutive processes. Both processes involve one group of people subjugating another; however, Said (1993) differentiates them by referring to discourses of imperialism as the practices, attitudes, and representations colonizers enacted in their own homeland, most often to initiate and justify their extended rule. On the other hand, he delineates discourses of colonialism as the material and ideological *expressions* of imperialism *within* the colonized territory. In his book *Postcolonialism*, Young (2001) more clearly delineates this home/abroad distinction, the historical relationship between colonialism and imperialism, and the form of power most closely associated with each type of domination. Therefore, while Said's work has influenced much of this project, I have adhered more closely to Young's conceptions of colonialism and imperialism.

Although colonial rule was variously enacted in different contexts (see Steinmetz 2002, 2003; Young 2001), I focus on colonialism as the pragmatic and diverse sets of *practices* used to establish and manage colonies abroad, as well as the "material conditions of the political rule of subjugated people by European colonial powers" (Young 2001, 27). I use the term imperialism, from a broadly Marxist perspective, to describe a system of global social, economic, political, and cultural domination that operates from the metropolitan center through an *ideology* of expanded and grandiose state power. Unlike the pragmatic and idiosyncratic business of colonialism, imperialism is driven by an ideological exercise of power through political and economic influence. This influence establishes a general system of domination over ostensibly inferior subaltern populations[10] by facilitating institutions and ideologies, *without* the political agenda of overt rule (see also Williams 1988). As Ashis Nandy (1983) argues, imperialism constitutes a general guiding principle that follows from colonialism, to make those previous practices culturally meaningful to expansionist states.

While we can use the term 'postcolonial' to depict the global political situation of nominally independent nation-states (with some important exceptions, such as British Gibraltar, French Guiana, and the Spanish Canary Islands), 'postimperialism' remains a misnomer, as systems of imperialism still operate throughout the contemporary world. Indeed, there is a consensus among the political Left that global power relations have not shifted in any significant, material way since the end of the colonial era. Since World War II, many Marxist-informed scholars have developed the concept 'neocolonial' to analyze the position of formerly colonized countries, which are largely unable to wrestle control of their economies from their colonial masters, despite political sovereignty. The concept of 'neocolonialism' usefully illustrates some enduring forms of colonial practice, but it has been primarily invoked to delineate continuing economic or military forms of power, without much attention to the history of cultural domination. Moreover, according to Young (2001, 48), the emphasis that has been placed "on a continuing neocolonial dominance has the disadvantage of suggesting a powerlessness and passivity which underestimates what has been achieved since independence, including the independence movements themselves, perpetuating stereotypes of helplessness even while it implies sympathy, and reinforcing assumptions of western hegemony with the third world being portrayed as its homogeneous eternal victim." With these problems in mind, I employ the term imperialism throughout my analysis, rather than neocolonialism, as it delineates a legacy of global domination into the postcolonial present, acknowledges a *cultural* historical inheritance (see Said 1993), and, as it emphasizes the conditions that both determine and undermine that global system, opens space for an analysis of subaltern resistance and agency within such conditions.

The term 'postcolonial' is one conceptual descriptor for this project. Following David Spurr (1993, 6) and Robert Young (2001, 2–11, 57–69), I understand 'postcolonial' to incorporate an element of praxis or activism, as well as history; it marks not only anticolonial struggles and the dismantling of colonial institutions, but also a search for alternatives to imperial discourses in the present. According to Young (2001, 58), "postcolonialism is both contestatory and committed toward political ideals of a transnational social justice. It attacks the status quo of hegemonic economic imperialism, and the history of colonialism and imperialism, but also signals an activist engagement with positive political positions and new forms of political identity." However, forging a set of discursive practices and political identities that resist imperialism in contemporary settings requires a prior understanding of their legacy of domination in the present. This project takes up the postcolonial challenge to engage with experiences of imperialism and their present effects at the local level of formerly colonized societies so that I can augment imaginings of alternative practices and more just social realities.

The objective of my project, then, is to develop what I will call a 'feminist sociology of imperialism' by temporally updating and empirically grounding previous feminist studies of colonial-era texts through an ethnographic study of how the subjectivities of Western women in present-day Gilgit are constituted through local and global relations of power as manifest in discourses of imperialism, gender, race, class, and sexuality. A social scientific study such as this affords women the opportunity to speak for themselves about how they experience discursive pressures in their everyday lives abroad. In addition to this epistemological consideration, this type of sociological analysis has important theoretical implications. Contemporary ethnographic evidence can augment what feminist historians, literary theorists, and geographers have established for colonial-era literature, and reveal the legacy of Western women's imperial involvement in South Asia. I frame a feminist sociology of imperialism, therefore, not as a specific system of principles or tenets, but rather as a field of study that employs sociological research tools to extend and enrich postcolonial understandings of the continuing relationship between Western women and imperialism.

By investigating how Western women in Gilgit enact various discourses, boundaries, and social orders of domination, and thus how they incite and resist social change, I aspire to contribute to a post-colonial understanding of transcultural power relations in South Asia, especially as they play out between local Muslims and Western non-Muslims. Theorizing these dynamics of power and resistance is now especially crucial, as we strive to understand and transform the transcultural discursive context of a post-September 11 world.

Parameters of Study

This project focuses on the lives of 30 Western women, most of whom are British, Canadian, Dutch, American, and Australian volunteers working for international development agencies in Gilgit on two-year contracts negotiated mainly through Volunteer Services Overseas (VSO). The majority are teachers, librarians, English-language coaches, and teacher trainers who instruct local educators on new methods of teaching the curriculum in English. The rest are health workers and partners of upper-level development personnel who became involved in development work after they arrived. Some women in these latter two groups have decided to settle in Gilgit for the long term. When my research participants have not followed their partners to Gilgit, they have chosen to travel there for reasons of self-determination, philanthropy, and job advancement, although most of them would have preferred work placements in a non-Muslim country. The experience of travel enables them to develop a sense of themselves as independent individuals—women who are more assured, competent, and authoritative than they were before they left home. In terms of philanthropy, most of them also come to initiate sociocultural reforms by revamping the local education system and 'freeing' Muslim women from an ostensibly oppressive Islamic culture through an unproblematic transfer of Western expertise and their own 'liberated' selves as examples. Working abroad empowers them by increasing their knowledge, specialization, and experience, which can translate into professional advancement, work autonomy, and pay increases once they return home. Teaching overseas in educational development also allows them to realize their intellectual potential and to garner some authority by training mostly male teachers, being Western educated, and representing their development work as an essential cultural 'improvement' project. In deciding to accept a job placement in Gilgit, Western women attempt to use the sense of authority they gain through travel and 'benevolent' development work to protect themselves against the gender oppression they expect to experience when living in an Islamic society. Consequently, my research participants have contradictory impacts on relations of power in this transcultural setting, impacts effected as they harness a range of global opportunities afforded by international development to refashion lives and identities that secure greater degrees of self-determination, but which simultaneously compromise the autonomy of Gilgitis.

In contrast to the portraits of Euro-American women in colonial India painted by Margaret Strobel (1991), Pat Barr (1976), and Marian Fowler (1987),[11] Western women in present-day Gilgit do not form a homogenous community with common interests. Different groups of women, with various friendships and enmities toward each other, coexist uneasily as sociocultural reformers, closeted missionaries, housewives, friends, housemates, and coworkers. There are significant internal tensions within this community, and thus diversity among Western women. For example, as I discuss in chapter four, the group of volunteers who lead less materially extravagant lives than long-term salaried expatriates develop an analysis of the latter group's imperial practices around the hiring and treatment of servants. This self-critique of contemporary imperialism unsettles dominant social relations between Westerners and indigenous people, as well as the personal relations among Western women.

Neither is this a permanent settlement. Rather, it is a temporary group of Western women, haphazardly thrown together by the accidental and limited circumstances of development job placement. The community is easily dismantled, fragile, and contingent.[17] Indeed, post-9/11, international development agencies recalled all volunteer workers from Pakistan, and shifted longer-term residents south where they would be closer to international airports for quick evacuation. Only since the fall of 2002 have volunteers begun trickling back to short-term placements in the Gilgit area. Although this community no longer physically exists as it did during the time of my research, my participants have undoubtedly taken away from Pakistan many life-altering experiences. As the opening quotations suggest, Western women are often stimulated and enriched by their experiences in Gilgit. Their time abroad produces powerful, persistent experiential effects and memories that dramatically shape subsequent self-understandings, principles, loyalties, and desires.

During the summers of 1999 and 2000 I conducted 37 in-depth interviews, nine months of participant observation, and several group interviews with Western women in Gilgit to explain how their subjectivities (and, to a lesser extent, those of the local people with whom they interact) are constituted in this sociospatial setting. I also kept a journal of my thoughts, experiences, and activities so I could examine my own part in perpetuating and resisting particular power relations, which drew me into the research as another participant.

Initially, I contacted a Canadian acquaintance in Gilgit, who joined my research project and introduced me to her foreign female friends and workmates. I then used a snowball sampling procedure until I accessed eight potential research participants. I described my research to these women as a study of their practical attempts to make themselves comfortable while they lived in Gilgit, including their daily routines, work and movement patterns, social interactions, and living arrangements. When they consented to participate, I organized a group interview to brainstorm pertinent research topics and interview questions. These eight women also suggested other participants and provided introductions. I was never certain of the exact number of possible participants, as work contracts wrapped up and women came and went. But after two summers, this first group of eight women furnished me with enough introductions to enlist 30 participants in the Gilgit region, seven of whom I followed during both summers.[13] In an attempt to 'describe,' but not reify this group of women, table 1.1 delineates eight facets of their subjectivities that emerged as central to their self-images and practices during our interactions.

Table 1.1   Facets of Subjectivity

Lifeline	participants are aged between 23 and 59; five are in their twenties, but most are between 37 and their early 50s; they usually identify as one of two 'life' groups: young (unmarried, no children, desirable) and seasoned ([once]married, with children, past desirability, and often, in their words, 'grandmotherish')
Ancestry	most participants, whether they consider Canada, the United States, Australia, or the United Kingdom 'home,' claim an Anglo-Saxon heritage; four said they had a Germanic or Nordic background
Formal Schooling	all but two participants are university educated—many have advanced degrees, teaching certificates, and English as a Second Language training
Vocation before Coming to Gilgit	one freelance photojournalist; one postal carrier; one business woman; one retired civil servant; two nurses; three housewives, two of whom are also part-time teachers; one doctor; one feminist lawyer—the rest are educators (teachers, teacher trainers, or education consultants)

Continued

Table 1.1   Continued

Religiosity	to eight women, Christianity is a fundamental component of their lives (five of them are missionaries)—the rest claim to be nonpracticing Christians or to have no religious affiliation
Station	they all self-identify as 'middle class'
Partnership	three lesbians (two of whom are partners); four are in long-term relationships with men, but are not married; eight are single; nine are married; five were once married
Progeny	14 participants have children

I might have been skeptical of Pierre Bourdieu's (1999, 615) claim that "conditions of discourse which might never have been spoken" can be generated through such intensive interactions if it had not happened during some of my own research experiences. Many of my participants recognized our social exchanges as exceptional opportunities not only to testify and explain themselves to me, but also to construct their own understandings and explain their lives to themselves. For example, Elena responded to my 'thank you' at the end of the interview by saying "*Pleasure*. It's very nice. It's actually wonderful talking about stuff, because it helps you sort it all out really. This has really helped me put my thoughts together and understand my life here better." Some participants also gained an "induced and accompanied self-analysis" (Bourdieu 1999, 615) through this process. During a painfully self-reflexive moment, Evelyn confessed that

> I feel my living here has been sort of walking around in a dream. Do I not notice all these things [that you've asked me about]? I felt like a failure after [our last interview]. And I feel after this one that the questions on friends hurt more than I thought they were going to hurt. Not because the *questions* hurt, no. Finding the answers was more painful than I thought. I thought 'Oh dear, have I really lived this long, not really feeling . . .' Everybody's *so nice*, that's not the question. *You know* that's not the question. But feeling that there's nobody who's special *for me*. That's hard. And what does that do to you? Does that make you shrivel up? And why are others so different?

Evelyn used this moment to work through painful responses to my questions and to consider how her actions might change. Thus, our interactions were a conduit and a construction site for participants' experiences, and the thinking processes prompted by my questions had some limited impact on processes of subjectivity formation. By inviting participants to consider what may have been left unthought, our interactions allowed them the opportunity to contemplate their lives, making it a reflexive exercise for both parties.

The analysis of subjectivity formation that results from this compilation of data allows me to develop a theory of agency that explains how these Western women in Gilgit are situated within local and global relations of power, especially as they operate through discourses of gender, race, class, sexuality, Orientalism, and imperialism. As my central point of inquiry is the formation of Western women's subjectivities, I am interested to explore how they perceive themselves and their behaviors, how they represent and interact with Gilgiti people, and what sociospatial boundaries they (re)construct between Self and Other. Their narratives of indigenous people's behaviors, as well as their understandings of the social context of Gilgit, are additional concerns. My focus, however, does not extend to the motivations, intentions, and interpretations of Gilgiti people and expatriate men. This project also does not include an examination of the effects of Western women's presence as narrated by Gilgiti people. I draw from my own understanding of local social conditions when that is relevant, but it is not an explicit focus of analysis. The study is restricted to describing and analyzing the discursive processes and consequences of subjectivity formation among Western women in Gilgit.

I have combined these social scientific methods into an ethnographic study, because ethnography, as a methodology, provides a framework for detailing the routine aspects of Western women's daily lives in Gilgit, so as to bring the ordinary stuff of everyday life—the practices usually taken for granted—into sharper focus, and to reveal how those activities have political significance through their implication in the exercise of power. Ethnography has also allowed me to explore how the experience of living in Gilgit has, in some areas, shaken my research participants' "ontological security," which Anthony Giddens (1984, 287) defines as "a sense of confidence or trust in the world as it appears to be," and how it has reinforced it in other areas. Both of these advantages demonstrate that ethnography becomes an important component of the critical practice of social science by producing knowledge grounded in daily life that questions normative social foundations, reveals systems of domination in everyday activities, and identifies avenues for social change as it connects daily life to fields of power.

In order to avoid a sense of unwavering certainty and omnipotence in this work, I have attempted to follow Liisa Malkki's (1997) understanding of ethnographic practice as situated 'witnessing,' as an alternative to 'objective' describing from a distance. Witnessing implies a specific researcher positioning of herself/himself as an *active* listener and *participant* observer, and it involves a responsibility to testify cautiously. Furthermore, it implies an epistemological 'mode of knowing' that is rigorous, but partial. The contradictory practices I highlight introduce fissures of doubt in my interpretations, which readers are encouraged to fill with their own explanations. These alternative interpretations may produce further imaginings of more just social practices and realities. The attention I pay to life events demonstrates how discursive structures are created and dismantled by Western women through social contestation and the voicing of competing cultural claims. I have also attempted to write polyphonic readings of power in different chapters that build on one another and expose the partiality of each of those readings, and to show how one site of the construction of inclusion/exclusion acts as the silent, but integral, background for others.

In combination with the results of ethnographic witnessing, I draw on feminist, poststructural, and postcolonial theory to argue that Western women's subjectivities are multiply, ambivalently, and relationally constituted through various discourses of power operating cross-culturally in Gilgit. These bodies of theory, in avoiding the Enlightenment ideal of subjectivity and binary thinking, allow me to circumvent two recurring problems within much social theory. First, the Enlightenment ideal of subjectivity—caricatured as an autonomous, irreducible human self—has been thoroughly interrogated by such well-known poststructural, postmodern, psychoanalytic, and black, lesbian, and postcolonial feminist critics[14] that I do not need to rehearse their arguments against it here. These authors, to varying degrees, replace the Enlightenment ideal of subjectivity with one that incorporates a multiply constituted, embodied, nonuniversal, nonessential subject who is embedded within multiple fields of power at both conscious and unconscious levels. Therefore, these frameworks can better account for the differences among subjects and their multiple and conflicted constitution.

Second, many theories that explore what makes us who we are and how we enact our positioning within particular social settings are dualistic in that they tend to emphasize the capacity either of human agents to create and perpetuate social life or of social structures to determine, limit, or constrain social actors. Whether they address small-scale interaction patterns between agents whose conduct is largely unaffected by structural constraints, or large-scale processes and institutions that 'determine' human action, these theories frame structure as a social realm analytically separate from processes of agency/subjectivity. Social life is split into two hierarchical levels of existence, making it difficult for theorists to conceptualize social change as dynamic processes involving macro-level structures *and* the locally situated actions of agents. Nondualistic theories of subjectivity, such as poststructural models that employ Michel Foucault's notion of discourse, enable this kind of nonhierarchical analysis by positing recursive relationships between structure and agency. In other words, structure and agency are understood not as mutually exclusive, externally imposed, and hierarchically arranged categories, but rather as dynamic, interwoven processes. Structure is recursively related to agency, constituting it and being at the same time constituted by it.

Foucault is not the only theorist interested to develop a recursive explanation of social life. Anthony Giddens (1979, 1984), for example, develops a structuration theory that replaces the concept of structure as a solely constraining force that stands apart from agents with the notion of the duality of structure: human agency constitutes social structure, but structure is also the medium of that constitution. He incorporates a human agent into social reproduction, but simultaneously acknowledges that this agent is not an autonomous actor creating society from scratch (Bryant and Jary 1991; Dean 1994; Kilminster 1991; Murdock 1997; Nelson 1999; Pile 1993; Ritzer 1992; Thompson 1984). While structuration theory is helpful in this regard, it retains traces of dualistic thinking and the ideal Enlightenment subject.

It does so, first, because Giddens relies on a coherent agent (Boyne 1991; Kilminster 1991; Murdock 1997).[15] Second, this agent is also a universal social actor who is not thoroughly positioned within contexts of power. Although he (1984, 83–92) argues that a social actor is 'situated' in time and space, he does not theorize how actors may be located differently depending on their gender, race, class, or sexuality. Therefore, Giddens perpetuates the concept of an abstract, universal subject. Third, this self-controlled agent, who acts according to its own principles and rational choices, appears fully formed out of an unspecified social setting (Kilminster 1991; Murdock 1997). Structuration theory describes the behaviors and characteristics of this agent, but it does not explain how this specific type of agent developed. Structural considerations seem to be irrelevant until after the knowledgeable[16] agent materializes. By relying on this type of agent, Giddens precludes a thorough recursiveness in his theory.

Pierre Bourdieu's (1977, 1990; Bourdieu and Wacquant 1992) recursive theory of practice is also problematic. He seeks to understand how patterned social practices are generated and regulated by setting social practice, structure (habitus and capital), and relations of power (field) in a dialectical relationship. Consequently, he posits a situated, active, and power-constituted subject. But this subject is unified and coherent[17] (Farnell 2000; McNay 1999, 2000). His theory of practice, therefore, cannot be used to theorize a conflictual, ambivalent, and multiply constituted social actor. Moreover, Bourdieu foregoes both an agential subject and embodied action by suggesting that dominant structural imperatives are mapped uncomplicatedly onto bodies to form bodily dispositions that reflect faithfully the conditions under which they were formed and by not specifying how dispositions incite practice (Alexander 1994; Garnham and Williams 1990; McCall 1992; McNay 1999, 2000; Scott 1990; Swartz 1997). Identities seem to be easily and flawlessly impressed upon subjects. There is no ambivalence or ambiguity or dissonance between structural positions and the way they are lived by social actors. Bourdieu thus overlooks the mistakes subjects often make in assessing possibilities and restraints and the contradictory expectations, aspirations, and motivations they simultaneously hold, mistakes and contradictions that challenge his theory of disposition acquisition and expression. Moreover, when he argues that habitus (structure) disposes social agents to act in certain ways, he undoes the recursiveness of his logic of practice. In my opinion, Michel Foucault more successfully avoids this type of dualistic thinking. He also forsakes the Enlightenment ideal of subjectivity more fully than Bourdieu and Giddens by presenting a recursive theory of social life that posits a multidimensional, embodied, active, and power-constituted subject.[18]

I refer to Foucault's (1977a, 1978, 1980) concept of discourse as a historically specific system of rules, sometimes called organizing principles of society, generated in particular institutional settings that structure how people think and act according to the needs of hierarchical forms of power. According to Foucault, individual societies generate a regime of discourses to identify and sanction 'true' assertions and 'normal' behaviors. They are acts of signification manifest in language, institutional practices, behaviors, objects, technologies, and concepts, and exercised throughout the social body as normalizing, embodied surveillance techniques to produce knowledge, truth, and modern subjects. In other words, discourses function, through disciplinary knowledge produced in and circulated from institutional bases, to constitute and govern individual subjects, human populations, and institutional practices.[19] By sanctioning 'true' assertions and 'normal' behaviors, and delineating what can and cannot be said, thought, meant, and done in a particular society, discourses are instruments (and effects) of power that provide frameworks for understanding the world. Discourse, then, can be conceived of as "a system of possibility for knowledge," because it provides the matrix of statements and practices that allow and sanction certain 'true' statements and certain 'normal' ways of acting and understanding (Philp 1985, 69). They thus provide the knowledge material from which individuals are fashioned and fashion themselves, make themselves socially intelligible in a particular context. What discourse makes unimaginable or unintelligible is socially marginalized. Discourses structure meaning, thought, and action in all realms of social life. But they are not unified or unchanging. There is a multitude or regime of competing, converging discourses circulating in every society, each relevant to a particular realm of social action and subject to challenge and transformation.

As subjects gradually begin to think and behave according to a particular set of discursive norms, they undergo the paradox of subject formation; they become subjected to power at the moment they become discernible subjects. Foucault (1980, 98) argues that "it is a prime effect of power that certain bodies, certain gestures, certain discourses, certain desires, come to be identified and constituted as individuals." Power, through discursive investments, produces subjectivity. And subjects, creating themselves out of discourses, are agential vehicles for that material, corporeal power. In this way Foucault's theories of discourse and subjectivation posit a recursive relationship between structure and agency, as well as an embodied, multiply constituted, and active subject.

In analytical terms, discourses comprise, animate, and arrange fields of meaning production—texts—and shape their interpretation. Texts in this context are not understood as finite written records of communicative acts (the common notion of the term), but rather as dynamic sets of signifying practices associated with all cultural productions, sites, and spaces for the discursive construction of meaning and distribution of knowledge (Barthes 1975). Texts, such as writings, maps, paintings, social interactions, and institutional practices, are the products of a discursive event when meaning is concretized or made material. By searching texts for the discourses that animate them, we can ascertain how meaning is achieved, how texts construct the things of which they 'speak,' and how interpretations are created as objective 'Truth' as opposed to biased, subjective musings.

As mutually informing sites of meaning production that conceptually frame subjects' actions and interpretations of the world, texts as discourse shape 'reality' through the authoritative stories they tell about what reality is, how it is organized, the way it has developed over time, what is important or valued in that reality, and what is not. Texts thus exemplify certain subjective, invested, and political versions of social life that are implicated in social power differences and dynamics and the particular patterns of behavior appropriate to that socially constructed 'reality.' Embodied practices, such as clothing choices, can be considered texts, because they are cultural productions, sets of discursively mediated cultural codes. Therefore, bodies, like written, spoken, and visual forms of signification, can also be analyzed to discern the systems of rules that organize them and mandate certain courses of action. The effort to pinpoint how particular significations are produced and become dominant in all of these discursive sites is essential as a basis for challenging practices of power that produce limited, often subjugating, versions of 'the way things are.'

I have employed an analytic approach that ascertains what discourses or systems of rules structure meaning in a text to (re)produce a particular version of reality (Foucault 1972, 233). I have attended to the surface regularities in narrative styles and devises, figures of speech, tropes and motifs, and categories of talk or action.[20] Repeated silences, gestures, events, statements given as true, and behaviors regarded as normal are also important in assessing what meaning rules organize this range of texts. The analysis of organizational rules is different from what many hermeneuticists call the 'interpretation' of a text in that it is not concerned to uncover the stable, original, and 'hidden' meaning buried within the text, or the author's intentions. Indeed, when texts are understood as fields of signification that produce a plurality of shifting meanings, the search for an author's 'true' and intended meaning is impossible. What matters, rather, is the internal structure of the text, the internal practices of signification. Texts involve prior interpretations in that they draw upon other texts that themselves are based on yet other texts. This intertextuality, or the embodiment of other cultural productions, is "the process whereby meaning is produced from text to text rather than, as it were, between text and world" (Barnes and Duncan 1992, 2). Consequently, texts are not defined in terms of authoring, but of reading: different readers may interpret the meaning of a text differently depending on how they connect it to both the context in which they read it and other cultural productions. The meanings of social actions and institutions are thus unstable due to their dependence on a range of different interpretations, which are informed or structured by prevalent discursive frameworks for understanding the world. So, while I am concerned about the accuracy of an experiential narrative and how experience is constructed, I am not interested in the 'truth' of that experience. Rather, I analyze the interpretive frameworks (discourses) that structure its internal meaning, together with the experiences and the knowledge of the self.

Interview data comprise a series of texts in the form of experiential narratives. How did I go about understanding the narrated experiences of my participants? In epistemological terms, an understanding of reality as discursively negotiated suggests that there are no unmediated, pristine experiences that ethnographers can access and know. Because "what is said about it augments the experience of it" (Foucault 1978, 22), experience is understood as discursively represented. As such, it cannot stand as the 'origin' of 'authentic' knowledge, as many standpoint epistemologists claim (see Hartsock 1996; MacKinnon 1982; Smith 1988; Stanley and Wise 1983). Rather, we employ conceptual structures to interpret our experience. People understand and describe their experiences using authoritative frameworks of thought and action that actually comprise the social realities and experiences they describe. So, experience is part of a discursive system, and not a "reflection of the real" (Scott 1992, 24). Experience is not only what individuals *have*. It is also the medium through which subjects are constituted. By analyzing my participants' experiences as discursively mediated, I can "understand the operations of complex and changing discursive processes by which identities are ascribed, resisted or embraced" (Scott 1992, 33).

I do not want to claim, as does Joan Scott (1992), that all experiences can be collapsed into discourse (see Kruks 2001; Stone-Mediatore 1998). Michel de Certeau (1984) argues that there are everyday practices and ideas (and I would add emotions) that have not been organized into discursive systems, and thus can incite new versions of reality. Experience may consist of practicing these unorganized elements. But, following de Certeau, I think discourse analysis can make us attentive to aspects of experience that are not organized by discourse and trace their potentially resistant effects. It allows me to understand embodied practices and spoken and written experiential narratives as power-filled discursive practices, and to discern how subjectivities are discursively constituted through multiple axes of power and how particular versions of reality are perpetuated and resisted through words, actions, and emotions.

My theoretical aim in exploring the interconnections among multiple axes of power and differentiation is not a new endeavor, even in the colonial context. But my particular study of the "politics of intersectionality" (Brah 1996, 16) can be beneficial, because each confluence tracing reveals new meanings and new ways to conceptualize the exercise of power, especially in postcolonial space. This book, therefore, can serve to enrich current understandings about these interrelationships.

In addition to making a contribution to current thinking about how discourses intersect and the way contemporary transcultural power relations and sociospatial boundaries are lived in South Asia, this study augments ongoing debates about how productively to conceive of the subject and subjectivity formation in both metropolitan and postcolonial settings. I also, but less explicitly, contribute to the literature on transnational flows of people and culture in the current context of globalization, and the effects of Western women's presence in globalized space on transnational power relations.[21] Finally, this project yields insights into the experiences of contemporary Western women living abroad, which have been largely neglected in feminist and globalization literatures. While scholars have offered guidelines for developing cultural sensitivity for Western women business managers working in Asia (Hebard 1996) and proper social and pedagogical training for teachers on their way to China to teach English (Boyle 2000), they have discussed only superficially American women's complicity in cross-cultural power arrangements in the Gulf states (Caesar 1991). Yeoh et al. (2000) and Yeoh and Khoo (1998) provide more satisfying analyses of expatriate women who have followed their spouses to new jobs in Singapore, find themselves reduced to economic dependency by immigration law, and negotiate their newly imposed domestic identity through volunteer community work. Deborah Mindry's (2001) analysis of the 'politics of virtue' and colonial discourses of benevolent motherhood ensconced in Western and local bourgeois women's philanthropic activities with 'oppressed Third World Women' in South African development agencies is also interesting and pertinent to discussions of global relations of power. But none of these studies discusses in detail the ways in which Western women are implicated in transcultural power relations as they live, work, and make for themselves a 'home away from home.'

Despite these benefits, contributions, and intentions, any ethnographic venture, especially one conducted in a postcolonial setting by a white Western researcher, is necessarily imbricated in the very power structures that are the subject of inquiry in this book, through the questions asked and the style, methodology, and concepts employed to develop explanations. Although this project emerged out of a personal interest in my own discursive position within this transcultural setting, by concentrating on metropolitan subjects, rather than Gilgitis, I was also responding to an "awareness of the role played in the study and representation of 'primitive' or less-developed non-Western societies by Western colonialism, the exploitation of dependence, the oppression of peasants, and the manipulation or management of native societies for imperial purposes" (Said 1989, 207). This anti-imperial sensitivity extends in my project to thinking the history of the metropolis together with the history of the (once)colonized and avoiding, as an 'outside' scholar, a visualization/representation of Gilgit for the Western gaze. I have attempted to turn the 'observing eye' at work in a visual ethnographic paradigm to the interplay of positioned 'statements' in a discursive one (Minh-ha 1989).

However, as no structure of knowledge can ever stand free of the social, cultural, historical, and political formations of its time, this ethnography is ultimately tied to imperialism in several ways. First, as a witnessing ethnographer I have a status, a field of activity, and a moving locus that mimics the imperialistic relation itself; imperial power has shaped the theoretical perspectives, conceptual frameworks, and standpoint from which I can witness and interpret in this contemporary ethnography. This enabling intersection between imperialism and ethnography is perhaps most notable in that imperialism makes the focus of ethnographic study accessible as it sustains the physical proximity between witnessing Westerner and participating non-Westerner, and thus helps achieve an often one-sided transcultural intimacy. Second, and relatedly, this ethnography is largely driven by Western theories and methodologies borrowed from phenomenology, symbolic interactionism, ethnomethodology, Marxism, feminism, and poststructuralism, despite my reliance on several influential transnational conceptual frameworks.[22] The notion of methodological reflexivity I employ, for example, is rooted in questioning the relationship between Western epistemologies and modern hermeneutics. Furthermore, the book perpetuates Western styles of discourse, and, in so doing, may recirculate images that feed into, connect with, and enhance political processes of dependency and domination.

Third, Empire has traditionally attempted to produce, acquire, subordinate, and settle space in its quest for colonial and imperial control. This ethnography is inculcated in that exercise of power by exerting some control over the geography of Gilgit, by enabling Western readers in some small way to 'know' it and its significant places, and not only through mapping exercises. And finally, I give accounts of my stay in the field, as well as my research participants' and my own institutional and material standpoints in the everyday world of Gilgit. However, these accounts are not *fully* connected and made problematic in relation to local realities, as I have mentioned. In framing my research questions according to postcolonial principles informed by colonial discourse analysis and postmodern and feminist ethnography, as well as to time and skill constraints, I have concentrated on the (re)circulation of global power at the expense of local voices that make their own sustained adversarial claims about imperialism. While there is no 'authentic native point of view' to recover, as Spivak (1988) argues, Gilgitis are largely overlooked as independent social agents in this project, although not as overtly as South Asian servants are in Raj-era British novels, making the book an indirect and ambivalent agent of political dominance. As Said (1989, 210) contends, "[t]o convert [subaltern subjects] into topics of discussion or fields of research is necessarily to change them into something fundamentally and constitutively different." So, the paradox that animates postcolonial studies more generally—to study the (once)colonized and risk their being judged, managed, controlled, regulated into Western social relations, or to focus on the global exercise of power and resistance at the expense of subaltern voices and experiences—infuses this text as well. While it seems that my postcolonial ethnography cannot escape the problems of ethnocentrism and imperialism, I have attempted to maintain a critical language and a critique of that language by naming some of the specific instances of discourses that construct and are perpetuated by my account of transcultural relations in Gilgit.

Tracking the Text

As an *ethnographic* study of Western women's subjectivity formation in *postcolonial* space, this book is self-consciously written as an opening, an invitation to further research that explores the "discursive formations, representations, and practices of power in contemporary social contexts, together with their relations to the colonial past," in order to imagine alternatives to colonial-era discourses (Young 1990, 175). My hope is that, in the chapters that follow, the multiple and conflicting dimensions of Western women's subjectivities will be evoked as complex, lived experiences, as materially practiced rather than abstractly imagined, and as historically situated, with varying meaning and effect, in a centuries-old transcultural field of power. Chapter one, which can be understood as a truncated "ethnoscape" (Appadurai, 1990) of Gilgit in the form of a narrative about the bazaar, further establishes the context of research and analysis by evoking the research setting, my position in the research process, and Western and indigenous women's daily lives. Included here is a brief and partial description of my research participants' perception of Gilgit and its citizens. These perceptions are historically and discursively linked in ambivalent ways to those of their Western predecessors. I trace these connections between present and past by outlining the history of transcultural interactions in the area and describing the metropolitan cultural worldviews that inform and frame those interactions. These worldviews are important, because they predispose Western women to understand Gilgit and Gilgitis in certain ways before they arrive in town and to behave once they settle there in a particular fashion. This contingent historical legacy, characterized as much by variation and rupture as by linear continuity into the present, provides further context for my analysis of subjectivity formation, to make the details of analysis more meaningful.

Chapters two, three, and four analyze significant processes, sites, boundaries, and consequences of Western women's subjectivity formation in Gilgit. Chapter two examines the mechanisms through which my research participants' sexually vulnerable subjectivities are forged ambivalently in relation to Gilgiti men. I pay particular attention to Western women's reactions to the 'lascivious male Other,' the sociospatial boundaries those practices sustain, and the spatial implications that arise from sexual boundaries of exclusion. Chapter three analyzes some of the processes by which Western women's 'liberated' subjectivities are formed in relation to their experiences of travel and to 'oppressed' Muslim women. I include discussions of my research participants' representations of Gilgiti women and of the importance of clothing both as a sign of local women's ostensible oppression and as a site of subjectivity formation for Western women. Moreover, I outline the relevance of development activities to their self-understandings. Chapter four analyzes Western women's Gilgiti homes as spaces of subjectivity formation and social exclusion. I examine the discourses women exercise and the practical efforts they take to create homes where they feel comfortable, relaxed, and in control. I also discuss how my research participants govern their servants and children and how those practices both shape domesticity and perpetuate racial and cultural exclusivity as components of imperial power. In the conclusion I use the theory of subjectivity formation outlined in the preceding three chapters to develop a theory of agency that explains how Western women are situated within local and global relations of power. By studying subjectivity formation among Western women in Gilgit in this way, I hope to begin the process of understanding how this contemporary transnational population is implicated in forms of imperialism and interconnected relations of power in postcolonial times.

Chapter One

Bazaar Situations

Daily excursions to the bazaar are customary for many Gilgiti men in nonwinter months. They go to buy perishable foods, household items, kerosene, and propane from China, as well as to chat with friends and associates over cold drinks or tea. Some families own vehicles, but most men get around town either on foot or by hailing a Suzuki van, the local vehicle used for public transportation.

That routine was no different for me. In contrast to many of my research participants, I eagerly anticipated this daily jaunt to the bazaar, not so much to shop as to be part of the bazaar activities, to view the interactions between people, to keep track of how shops and specific bazaars have transformed, to visit Mustafa and his children over cold drinks in his rug and jewelry shop, and to enjoy the bazaar's culinary delights. Elena expresses a similar enthusiasm: "We love looking in all the shops, honest. You find new things there all the time, you just have to look around. Usually we go to the bakeries and the fruit and veg shops. And there's a fantastic silk shop down by the post office. We go there *a lot.*" But most Western women who live in Gilgit find visits to the bazaar a chore at best. René, for example, comments: "I avoid the bazaar at all costs. To me, going to the bazaar is something I hate. There's no fun in it at all. You know, you don't get to explore . . . So I go to the bazaar, in the car, and I come home as fast as I can . . . I'm ready to be defensive. I'm really very much on my guard." Abbie tries to "only go once a week, but it's often more like twice. But I don't like going to the bazaar, I really don't enjoy it. I just do it because you have to. It's just so tiring, and it's not fun. I drive there and I go to Marshall's and to one or two *subzi*[1] places, and that's about it." Andy and Janet prefer to do their shopping from inside an office vehicle on the way home from work. Andy says that the vehicle helps her feel more comfortable

because I feel contained, safe, in place, and everyone understands white women in a vehicle . . . If I had someone who lived near me, I would go with them to the bazaar. But I'm not going to go there by myself. I sometimes go there by myself if I have to, but I wear my walkman to switch off . . . I can manage to get a Suzuki ride and then do some shopping. But there's no way I'm going to walk down to NLI market by myself, because you have to, it's just, I mean, not least because it's so *hot* at the moment, but . . . I'd love to go to the bazaar and go window shopping and not feel stressed. But I *don't* feel that way. I feel stressed by it. So I rely heavily on the vehicle.

I understand why many of my participants try to avoid this public place. The bazaar is crowded, hot, dusty, noisy, and stressful, and it is an overwhelmingly male space. Western women wandering there attract prolonged stares and, less frequently, other forms of unwanted attention from some Gilgiti men. A few women perceive this staring as an exaggerated, but harmless, form of curiosity. For example, Amanda thinks that "men stare because they don't see women normally. So they always look. And it doesn't matter, I think." Others are more ambivalent about it or find the staring overtly hostile. Elena, for instance, claims "It's curiosity, mostly. I mean, it may not be, but that's what I always tell myself. It's just curiosity . . . And I don't notice it anymore because I just don't look. I just avoid people." According to Janet, "it's this blank, unsmiling sort of staring which looks disapproving. And I realize that part of it is that you are a strange animal, a lone woman on the street, but then there's this, what seems like just a complete *contempt* . . . Some days it does, it feels like hostility and contempt, just contempt for me as a person." But whether Western women are enthusiastic, ambivalent, or distressed about the bazaar, it is certainly one of the most significant spaces of subjectivity formation for them in Gilgit. Their practical efforts to situate themselves in this space and to negotiate how they will behave, dress, and interact with people there contribute to not only their identities as white, Western, privileged, desirable/desiring women, but also the identities of present Gilgiti men and largely invisible Gilgiti women.

It took me several extended visits to Gilgit before I noticed local women in the bazaar. My senses were initially too overwhelmed by the unfamiliar buildings, people, clothing, interactions, shops, smells, and languages[2] to see them. My early Orientalist perspective on Muslim women also disposed me to not look for them. But over time, after I had come to know several shopkeepers and Gilgiti families and they began to hail me in the streets, I started to search the crowd for familiar, friendly faces. And once I looked for specific people, I began to notice the details of the bazaar, including the presence of women. Women do not fill the streets of Gilgit by any means, but they are there in small numbers, mostly driving to and from work in development agency vehicles, shopping out of family cars, strolling around sheltered bazaars wearing various forms of purdah[3] or rushing from one *het*[4] to another across busy bazaar streets so as not to be noticed by nearby men.

Coming to notice and comprehend some aspects of the lives of Gilgiti women was a long learning process for me. My first exposure to the worlds behind the bazaar was through Mustafa, who invited me and my partner to his home for an evening meal very shortly after we met him in his shop in 1986. Mustafa's wife, Uma, is a Sunni who observes strict purdah (against her husband's wishes), so initially only I could freely roam the house with the rest of the family.[5] With David there, Mustafa's children brought the food to the guest room where everyone but Uma and the youngest baby ate. Although I did not know Shina, the predominant language of Gilgit, or much Urdu, the official national language of Pakisan, Uma, over time, taught me how to operate in a Gilgiti kitchen, took me on jaunts around the *het*, and introduced me to relatives in her father's house. It was by wandering the narrow paths through the neighborhoods behind the bazaar that I first got a glimpse into the lives of Gilgiti women, who seemed to move freely and uncovered there, visiting friends and family and buying household supplies at the ladies' shops.

That glimpse was broadened in 1995 and 2000 when I, with my partner and daughter, lived with two other Gilgiti families for extended periods of time. These people are Ismailis from Hunza to the north of Gilgit, followers of the Aga Khan. None of the women in these households observe purdah, although they often cover their heads while driving through the bazaar in case their more conservative relatives or neighbors should happen to spot them and make their apparent 'lack of honor' a community issue. At the time we lived with their families, the women in these families were either at medical school in Lahore or agricultural college in Peshawar, working at the Aga Khan Rural Support Program office in town, or studying to pass their matriculation exams. They taught me how to wear *shalwar kameez*[6] and how to cook in the local style. They also explained the differences between Sunni, Shia, and Ismaili Muslims in the area, as well as how they tried to avoid getting married after their years of study. I commiserated with the working women about the difficulties of combining the responsibilities of home and children with the demands of the office and the effects of those time demands on our relationships with our partners.

We also discussed how the growth of the commercial centers of Gilgit, the spread of bazaars into residential areas, and the increasing numbers of nonfamilial men walking the small paths of the *het* to and from the bazaar have radically altered local women's lives. As these men peer into household compounds from the third storey of hotels, restaurants, and office buildings, women are forced to practice purdah inside their neighborhoods and even inside their own homes (Gratz 1997, 494). Roof-top activities and unveiled walks to visit friends and family are no longer possible, but higher walls around gardens and houses can offer some protection from the unwanted male gaze. Katrin Gratz (494) explains that migrant women from other parts of Pakistan often suffer the greatest loss of mobility and freedom, because they have no relatives around them, and therefore have no 'legitimate' reason to leave their household compounds. The severe isolation and immobility some Gilgiti women experience can lead to serious mental and physical health problems.

Although 'women's place' is shrinking in many parts of town, Gratz explains that Gilgiti women have developed spatial strategies to enlarge their sphere of activity. First, affluent women incorporate vehicles into gendered space as a 'moving house' driven by male relatives that transports them through male space. As an extension of home, Gilgiti women do not feel the need to veil themselves when they are inside the family vehicle. Second, poorer women rely on hiring private Suzuki vans to bridge the male space between the *het* and public facilities such as schools and the hospital. Women veil to board Suzukis, but once inside, protected by the curtains that enclose the passenger seating area, they unveil themselves to watch the activities in the bazaar. Mustafa explained the laughter emanating from one enclosed Suzuki in the bazaar as "Ladies' picnic day!"

Gilgiti women also enhance their sphere of economic activity through ladies' shops. These shops, which are run by and exclusively for women according to seclusion rules, are housed in private homes or rented rooms in the neighborhood. The first ladies' shop opened in 1960, and it remained the only one for approximately 15 years. However, since 1984, about 35 new shops have opened throughout Gilgit, offering cloth, shoes, baby clothes, school uniforms, make-up, jewelry, and craft items for sale. This rise in the number of ladies' shops indicates that local women's education and economic situation has improved, that there is a growing female labor market in Gilgit, and that there are specific consumer demands in town which only Gilgiti women shopkeepers can meet.

If Gilgiti women's lives are as varied, rich, complex, and challenging as I describe, then why do so many of my research participants view them otherwise? Most Western women in my study have an ambivalent view of indigenous women. While all of them acknowledge that some women with the appropriate education, class status, and religious background have the latitude to become teachers, development workers, and Lady Health Visitors,[7] many of them retain the notion that Gilgiti women are absolutely oppressed by the dictates of Islam and male family members. These Western women, like many memsahibs in colonial India, often perceive indigenous women as passive beings who rarely even recognize their subordinate position and thus do not think about resisting in any meaningful way what the former define as Muslim women's oppression. For example, many of my research participants who have either met or heard of Uma blame her 'restricted' life on Mustafa who, they argue, enforces Sunni rules of purdah on her while he lives an emancipated life among the foreigners in town. Andy admits that "someone like Mustafa I struggle with more than with most [Gilgiti] people here, because he has this completely liberated, Western approach, 'I drink, I smoke, blah, blah, but *my* wife stays in the kitchen' . . . it's fine for the men [to swim and go to parties] but not the women. And when you see men in Western clothes and their wives in burqa,[8] I won't go along with that at all. I really hate it." None of the Western women see Uma as an agent who chooses this kind of life for herself, despite being married to a man who implores her to develop a romantic 'love marriage' with him. Janneke feels sorry for Gilgiti women like Uma, especially when

> They have to sit in another room and be segregated . . . they served tea. All the men and I got tea, but none of the women . . . I felt sorry for the women. I'm sure they didn't think anything of it, but I did. For them it was so natural, being brought up in the Muslim ways, but I saw it as oppression. Western society is always stressing human rights, but for the local women, I wonder if they really see their inequality and subordination.

I, too, felt sorry for Gilgiti women during my first trip to Pakistan in 1986. I had developed a popular interest in (Euro-American) feminism, and this interest sensitized me to the absence of women in the street and to the veiled state of those who were present. This cloistering and covering of women offended, confused, and irritated my metropolitan sensibilities. I was vexed by my sudden conspicuousness, but, at the same time, I saw them as much more oppressed than any Western woman I could imagine. Before I became friends with women like Uma and developed a more nuanced understanding of their everyday lives, I identified Islam as the primary oppressor of these poor women, together with their husbands who were explicit collaborators in this system of complete female subordination.

Although she has not developed close relationships with Gilgiti women, Anguita sees things in a more nuanced manner, drawing on her own experiences and those of her friends at 'home.' She is frustrated that many "Western women think that women here are at home, repressed. I think that women are repressed in every home in the world. I have had Western friends, it would *embarrass* me to describe their lives . . . Part of the problem is these Western women come and think *they* are free." But while Anguita perceives gender oppression as a global phenomenon, Marion claims she has nothing in common with Gilgiti women: "And here in particular, in the way this society is structured, I don't think women here have a lot in common with women who come from my culture. So I hadn't expected to establish close relationships with women here." She remembers her conversations with the teachers with whom she works about problems of combining work and family life, problems women have when their partners are only "sometimes good," the hated *dupatta*, in-law difficulties, and women's lack of economic independence. But even while she notes these links with her own life, Marion eventually concludes that Gilgiti women are absolutely different: "Oh god, I wouldn't change places with them for anything, not for *anything*." Amanda agrees that she

> could never get warm with women from here . . . I had to fight for my rights at home, so [Gilgiti] women can fight for their own rights . . . Here it is much more necessary, because here there's so much injustice. You have to stay at home, and women here are always suffering . . . They give too many children birth. And they let men handle them like weak people. That's their lot. They can't be seen or heard. That's a big problem here for me I think, the plight of Pakistani women.

With respect to a solution, Amanda suggests "they should fight back, not just take it any more. But the women here, I think they see themselves as victims, and this I cannot understand. And it's not improving or changing."

René would improve the lot of Gilgiti women by giving them jobs: "The reason why I say that work would be a good thing for local women is that underlying that, I'm assuming that as they work they learn how to think . . . how to problem solve, how to think of cause and effect and sequence. This happens first, second, third, and if I do that then that will happen. You know, most of these people don't have that type of critical thinking." Lyn would "send lots and lots of [Gilgiti] women overseas so they can see that being a free woman doesn't mean being non-Muslim. I mean, I'm *sure* you can be a Muslim woman and still have freedom. At least I think you can, I don't know." Anguita insists that Gilgiti women do indeed have negotiating space within the oppressions they experience:

> Local women have a different *platform* of power; the house and children are in their hands . . . They have a say in the money, they have a say in the education, they have a say in the marriage of their children. So they run their platform and men run the outside platform, the bazaar life, and bring the money home. It's not equal, but they are *active*, they *have* a say . . . OK, I know because I read the newspaper that there is *abuse*, but, you know, I did in England four months in a women's refuge being abused by a Western man. So, a bastard misogynist is a bastard misogynist. There is no national or religious link.

If most of my research participants admit that they are not intimately familiar with the lives of Gilgiti women, then how do they come to have such strong preconceptions about what indigenous women and their lives are like? To begin to answer that question, which is, after all, an important part of my overall project, I trace the prevalent discourses about this place and its inhabitants that have been circulated by their Western predecessors.

Cited Transcultural Interactions

> The man rasped, "The people [in Pakistan] lie a lot. I don't know what to believe . . . Maybe we Americans have to stop being so naive." Naive? He did not strike Zareen as naive. She felt he had come with preconceived notions about Pakistan and intended only to reinforce them. (Sidhwa 1993, 175–176)

Exploration Sites

Frederic Drew was one of the first Europeans to write about his imperial adventures in what is now northern Pakistan. As he conducted geological surveys of Kashmir's mineral deposits for the British Raj, he recorded his impressions of local people for metropolitan edification: "The Hunzakuts are a dreadfully dirty people, far more so than any other tribe I have ever met with; their faces are blotched with black dirt, which they never think of removing. As a means of purifying, instead of washing, they burn twigs of pencil-cedar, and let the smoke and the scent from it come over them and inside their clothes" (1875, 431). Drew is not alone in assuming Western superiority when speaking about South Asian indigenous people.[9] For example, Durand (1899) portrays indigenous people in ways similar to Drew, Knight (1893), and Biddulph (1880). However, it is by infantilizing indigenous people that Durand expresses his patronizing, derogatory criticisms. For instance, he tells readers that the people of the Northern frontier have only childlike farming and construction techniques to make their desert environment more hospitable:

> It must be remembered that the people have no proper tools, no crowbars and dynamite to assist them. A tiny pick of soft iron, which looks like a child's garden tool, shovels fashioned out of wood, and a few poles as levers, are all that they have to work with. The use of mortar is unknown; all their walls are of dry masonry. Yet with all these disadvantages, with nothing but their eye as a guide to levels, they have carried out this great irrigation channel for six miles, and turned an arid desert into a garden. (162)

Although Durand acknowledges farming constraints and skills, he compares indigenous construction implements to children's garden tools. His evaluation of 'disadvantaged' Hunza building practices also represents them as lacking, as inadequate when compared to European construction techniques.

These early representations of the Northern Areas gained a more significant cultural cachet as they were selectively and divergently incorporated into the writings of later European explorers, travelers, scientists, and colonial administrators, who roamed the Karakorum valleys to add empirical detail to mid-nineteenth-century accounts of the physical and social contours of the area. Emily Lorimer and Jenny Visser-Hooft are two early-twentieth-century travelers whose writings about the Northern frontier solidify, but also partially destabilize, some of these authoritative representations.[10]

Lorimer lived in Hunza in 1934 with her husband Lieutenant-Colonel David L. R. Lorimer, who was a former political agent in Gilgit. They were both keen linguists, but Emily, observing European norms of appropriate femininity, wrote as many pages describing the indigenous people and customs in a maternal tone as she did outlining linguistic observations.[11] For example, she notes that "these illiterate people, who had never seen a picture in their lives, were able to recognize and name the tiny reproductions of themselves" (1939, 176). A patronizing attitude accompanies her sympathetic, feminized representations; she condescends to indigenous Others. For example, she alludes to the inferior status of indigenous women in this society when she acknowledges that "What the Nagyr women and older girls look like I cannot say, for when they sight a man they crouch down with their backs turned and either pull a cloth over their faces or bury their heads in their arms. Give me the Maulai [Ismaili] variety of Islam every time!" (277). As did Knight (493–494) and Biddulph (30) in an earlier period, Lorimer mocks local Shia practices, and approves of a more European variety of Islam (Ismaili). She implies a widespread inferiority of indigenous people, including inactivity and laziness, in her description of a baffling, substandard, and essentially different 'East' that contrasts unfavorably with the 'West': "Such characteristics are inexplicable to the European, and especially to the energetic Anglo-Saxon; the desire to explore and push into the unknown being innate in our race. As Rudyard Kipling has aptly expressed it, 'East is East and West is West, and ne'er the twain shall meet'" (1939, 51). This point of view appears to have gained authority by the 1930s, the beginning of the end of Empire: Lorimer not only uses stronger language than her predecessors did to express the 'strangeness' of 'Orientals,' but also cites Kipling, by then seen as an eminent authority on the Northern frontier of the Raj, as ideological support.

Jenny Visser-Hooft, a Dutch scientist who explored Karakorum glaciers with her German husband, condescends to local inhabitants by metaphorically converting them into beasts. She complains that she needed help in "controlling" the porters they hired to carry their gear through the mountains. "Testing their quality" was essential before embarking on any lengthy trip, and she notes that the ragtag porters looked like a "row of miserable sparrows" when they were asleep outside (1926, 24). But, like Lorimer, Visser-Hooft periodically adopts a tone of moral concern for the state of 'natives.' Western women's concern for indigenous inhabitants contrasts with men's colonial representations, which usually either ignore local citizens or refer to their simple, ignorant, and barbaric ways of life.[12] But feminized and colonial depictions often coexist. For instance, Visser-Hooft expresses concern over the working conditions of her porters. She notes that "It is no light matter even for men accustomed to the rough life of the mountains to carry a load of 40–50 pounds 7,000 feet up a barren, precipitous slope, exposed to the fierce rays of the sun" (185). However, in the same paragraph she abandons her concern by complaining about the inferior quality of these beasts of burden: "We could not grumble too much even if we had to go to bed without our supper because [our porters] had been so slow" (185).

The ostensible moral high ground from which these writers speak provides a position from which to judge the status of women in this colonized region. Visser-Hooft regretfully notes that Shimshali women seem always to be harassed by their husbands who issue mandates, often in the form of orders to stay indoors and out of sight. As she, like many Western feminists, condemns the station of indigenous women, Visser-Hooft simultaneously insinuates that she and her country-woman are free from these patriarchal oppressions. Perhaps her per-ceived position of power in this imperial setting allows her to disregard the gender oppression middle-class European women expe-rienced at home, such as the difficulties they faced being admitted to universities and getting well-paid employment or permission to travel alone. Her views about the status of indigenous women reaffirm her imperialist assumption that "Western society is best and that societies can be judged on their level of civilization according to the degree they approximate the idealized Western treatment of women" (Mills 1994, 41; see also Hall 2004; Levine 2004a; Wilson 2004).[13]

Postpartition Travel Sightings

Many of these ambivalent depictions of Gilgiti people, through a series of discursive ruptures and recuperations, survived even after the area was liberated from colonial rule in 1947, owing to the powerful ten-dency of later travelers, journalists, and scientists to cite in their writ-ings as authoritative the knowledge about the area that began to be encoded a century earlier. This citational practice is the process by which metropolitan travel experiences, observations, and understand-ings are ritually repeated over time. Ken MacDonald (1998, 229, my emphasis) argues that

> Like authoritative traditions, [travel experiences] are transmitted, with minor modifications, from one generation of traveler to the next. It is in the form of invented tradition and transmission that they take on a history, what we might call a *genealogy of travel* that links contemporary repre-sentations of people and place to those of the past. It is also important to understand that these representations do not originate with one specific individual or narrative but result from the conflation of discursive forma-tions with which the narrator is engaged. These representations must be understood largely within the contexts of their circulation and reception.

David Butz (1993) notes that the tradition of representing colonized people in derogatory ways may become less extreme and more insidious in Western travel accounts over time, as travelers try to find the 'good' as well as the 'bad' about the people and places they visit. But travelers continue to judge indigenous people by imperialist criteria. And even "the good [remains] as much for the West as the bad; it is good *for* the visitor" (Butz 1993, 162).

Around the time of partition, European presence in the Northern frontier faded as the British vacated the area and the newly emerging Pakistani nation came to be at war with India over Kashmir, a situation that made travel, in addition to local life, difficult and dangerous. Foreign travelers, journalists, and researchers gradually made their way back to the Northern Areas about a decade after Pakistan's Independence, to write more about the area.[14] Travel writers, in particular, have variously participated in the tradition of representing the Northern Areas to metropolitan audiences. Jewel Hatcher Henrickson, an American missionary working in a church-run hospital in Karachi, wrote *Holiday in Hunza* (1960) to detail her September vacation with the Mir of Hunza. She constructs a safe feminine persona for herself in her travelogue, especially by adopting a protective, maternal attitude toward indigenous people. She nurtures young women to become competent nurses, describes local child-rearing methods, teaches American games and religious songs to children, and expresses concern over the loss of local community members, property, and resources. Despite this moral concern for indigenous people, Henrickson retains a belief in the superiority of the one 'true' civilization and religion. She dreams of hearing mullahs chanting from the Holy Scriptures rather than the Quran, and is adamant that any new hospital established under her organization include a mandate for all Muslim patients to convert to Christianity. Combined with complaints about locals' inability to handle such simple tasks as answering the telephone and the 'primitive' facilities and working environments in local hospitals, her conversion fantasies represent all things Pakistani as falling short of a superior Euro-American standard.

In *Where the Indus Is Young* (1977) Dervla Murphy, an Irish professional travel writer, records her 1974 tour of Baltistan with her six-year-old daughter. Murphy foregrounds her maternity, not in regard to indigenous people, but toward her daughter. She stresses her "maternal rage," "maternal instinct," and "maternal duty," and reassures readers that she "unfailingly reads Rachel a bedtime story even in the most unlikely circumstances" (48). Her moral concern for 'oppressed' indigenous women is expressed by challenging child marriage and purdah, women's limited spatial mobility and 'endless' fecundity, and girls' inadequate educational opportunities. She claims that "In Baltistan one gets the general impression that women have the status of talking animals" (252). Despite these vague overtones of concern, Murphy recuperates imperial visions of indigenous people when she notes that "even those who speak English (of a sort) are quite ignorant about the greater outside world" (34). Condescending from her superior vantage point, Murphy criticizes "illiterate, gullible, hide-bound peasants" (12) and their inherent inability to "Westernize," and deplores their inferior educational system: "here Gol's scholars sit in rows imbibing what passes locally for education . . . possibly it makes its pupils literate in Urdu, but even this seems doubtful" (161). The "universally filthy" Baltis are no better than the Gilgitis, who are without "vigor, charm, and intelligence; I suspect their average I.Q. is rather below normal" (33). Murphy also refers to various Baltis as wasps and squirrels, and catches herself thinking that "there was perhaps more in common between [Baltis] and their animals than between them and us" (253).

In her award-winning travel book *The Golden Peak: Travels in Northern Pakistan* (1993) Kathleen Jamie often represents people and places in the Northern Areas in a more sensitive way than does Murphy. However, this sensitivity is uncomfortably juxtaposed with traditional Othering representations that *subtly* locate indigenous people in an unchanging past, characterize them as childlike, and equate them with animals. For example, Jamie observes that "Her clothes were full and modest, her hair tidily drawn back in a plait beneath a black *dupatta*. For a moment I forgot I was in Pakistan and thought 'Victorian'" (18). And methods for remembering 'strange' local names evoke imperial discourses of beastialization: "At first, when all the unfamiliar names confused me, I thought of Jamila as 'the leopard woman,' because of her outfit, her fast, leaner way of moving, and the dark mischief in her eyes" (19). Jamie, like many of the postpartition women travel writers before her, disrupts the imperial discursive legacy about the Northern Areas by representing indigenous people as three-dimensional subjects whom she admires. But she also perpetuates certain imperial visions of the region, especially as her famous travel journal becomes a 'must read' for the increasing number of foreign tourists who now holiday there.

Tourist Sights

Although travel narratives constitute a salient point of contact between Self and Other (Chaudhuri 2002), they, together with political histories of the region,[15] are only part of the 'must read' list for contemporary travelers to the Northern Areas. Tania Cobham (1997) has estimated that approximately 79 percent of these tourists use travel guidebooks for information and advice about traveling in the region.[16] Tania Dolphin (2000) argues that guidebooks, in concert with exploration and travel accounts, recuperate (as well as modify) some imperial representations of Northern Areas people, especially porters.[17]

Drawing on European travel lore, guidebook authors frequently represent indigenous men in the portering industry as a homogeneous group whose essential characteristics—especially their 'natural' inclination toward avarice, obstinacy, and inconsistency—become apparent through their interactions with foreign travelers. Isobel Shaw and her son declare that "Nagyr porters are expensive and difficult" (1993, 301). Mock and O'Neil frame their advice assuming the same defiance: "Trekking parties used to begin from Nagyr, but because of difficulties with Nagyr porters, they now begin from Baltistan. However, Nagyr porters are now eager for work and welcome trekkers to do this route from west to east . . . We have been told Nagyr porters promise to behave!" (1996a, 298). These porters gained notoriety in imperial literature for their 'difficult' behavior (thefts, lies, strikes, complaints), which was attributed to their inferior racial characteristics. For example, Breman (1989) tells how colonial explorers blamed porters' stupidity and incompetence on their Balti blood. MacDonald (1998) and Butz (1995a) argue, by contrast, that 'uncivilized' porter behavior can be understood as acts of resistance directed against foreigners who treat them inequitably by not paying them in full or treating them as subhuman. Porters' refusals to move, for example, may not have been manifestations of their 'unreliable and belligerent character,' but rather attempts to assert their will against *angrezi* who treated them like animals. But guidebook authors perpetuate a racialized representation of porters when they foreground porters' 'essential' characteristics and do not address the contexts of labor relations within which 'difficulties' arise.

Some guidebooks are also notorious for sustaining the imperial fear of colonized men's unbounded sexuality, which is based primarily on their ostensible 'primitivism' (McClintock 1995; Stoler 1997). Travel lore has it that colonized subjects' penises are especially aroused by the sight of white women. Guidebook authors recuperate the mythical danger of the "Black Peril" (Stoler 1997) when they consistently warn Western women of the dangers of traveling amongst Pakistani men.[18] Although these authors are too subtle to link 'primitive' men explicitly with a 'primitive' or heightened sexuality, this is just the assumption they make when they give the impression that Western women should be cautious of all Pakistani men. For example, John King, like Shaw (1990, 40–41) and Mock and O'Neil (1996a, 62–63), explains that

> Every year many women travel around the country solo and love it; others, even with a male companion, have been, in the words of one, "spat at, grabbed, grappled with, sexually assaulted, ridiculed and derided, shoulder barged and . . . made to feel most unwelcome and uncomfortable" . . . many men are isolated from what Westerners consider normal interactions with women outside the family . . . Thus women traveling on their own may be viewed as misfits or on the make . . . It's a tightrope act for a woman to get familiar enough with a local man to learn about his life or his village, while keeping him from interpreting the conversation as an invitation and turning up at her door later. Physical signs can help or hinder; follow local practice of not shaking hands with a new male friend, and keep eye contact to a minimum. (1993, 76–77)

King may allude to the variability of women's experiences with Pakistani men, but, by subtly characterizing "many" of these men as unsophisticated villagers dissociated from and not bound by cosmopolitan norms of appropriate sexual interaction, he implies that foreign women need to be ever cautious of these uncultured, sexually forward, and unpredictable folk. King transmits what MacDonald calls an imperial history of travel that links representations of contemporary Pakistani men to those in past travel accounts. By perpetuating the notion that Pakistani men pose a disproportionate sexual threat to white women, he infuses men's bodies with women's sexual fears, thus requiring a heightened ethical and sexual surveillance of Pakistani men by Western men and women. And this vigilance, in turn, helps to maintain the sexual, social, and spatial boundaries between locals and Europeans that were established in the imperial era (McClintock 1995; Stoler 1997).

Development Sites

All but two of my research participants either work for one of the numerous development agencies working out of Gilgit or are married to men who occupy the top positions of authority in them. And most of these expatriates become familiarized with Pakistan before they leave home, not only through the travel, exploration, and tourist books they have read, but also through reading material distributed by the development agency offices in their home countries. Every British volunteer who comes to work in Gilgit through VSO writes a report at the end of her two-year term. This report—in which volunteers describe their work experiences, relate their feelings about the bazaar and indigenous people, render judgments, and give advice to those volunteers who will follow them to Gilgit—is then housed in a library at the VSO head office in London. Many of my research participants prepared themselves for their stint in the Northern Areas by reading some of these reports, although they admit they would have read more if the material had not been so repetitive.

Like VSO, the Canadian government provides its development consultants with practical information about living in Pakistan. The Briefing Center (1986, Section 11.5) advises Canadian International Development Agency workers that

> In general servants are a must in Pakistan. The size of houses, the weather and local customs all make their employment desirable. If chosen well, they will give good service; most are dependable, honest, and hard working. It is essential to supervise them and give them precise, clear instructions . . . It's a good idea not to put temptation in their way. Basically honest, they have such a poor lifestyle it's unfair to leave valuables lying around . . . Most [servants] are fairly specialized in what they can and are willing to do.

Consultants are also advised to bring with them from home "4 loaf tins (you'll want home made bread)," "large stainless steel bowls (to make bread in)," and a "glass lemon squeezer (servants are hard on plastics)" (1986, Section 11.3). Authors not only imply that Pakistani bread is not to metropolitan liking, but also represent Pakistanis as ultimately untrustworthy, morally weak, incompetent inferiors who need constant supervision by their betters. And, like porters, these servants are presumed to be a natural and willing laboring class for wealthy expatriates.

The Canadian Briefing Center's (1986) advice extends to consultants beyond the realm of the practical and into the cultural; it provides a list of preparatory readings and viewings. Recommended readings include travel guidebooks to Pakistan, as well as Paul Scott's [1966] (1998) *The Raj Quartet*, M. M. Kaye's (1997) *The Far Pavilions*, E. M. Forster's [1924] (1976) *A Passage to India*, and Ruth Prawer Jhabvala's (1975) *Heat and Dust*. All of these Raj nostalgia novels, which Robina Mohammad (1999, 221) associates with a "neo-conservative remythification of the imperial past," reflect concerns about increased South Asian immigration to Britain in the 1950s and 1960s.

Margaret Thatcher's fear that "the country might be swamped by people with a different culture" (as cited in Brah 1996, 37) is conveyed in these novels and their 1980s movie adaptations. The 1984 *Jewel in the Crown* series, which was originally produced for Granada Television in Britain, is perhaps the most famous of these films. All of my British research participants are familiar with the characters, plot lines, and scenery depicted in the series.

Yasmin Jiwani (1992) argues that these particular novels and films circulate many of the images found in Raj literature and colonial missionary reports. Collectively they portray South Asia as a land of excess: excess poverty, disease, fertility, sexuality, violence, mysticism, disorder, and backwardness. In concert with European fears of the Black Peril, South Asian men are represented in these cultural artifacts as lascivious perverts obsessed with a desire for sexual intercourse with white women. This mythical obsession is animated in *A Passage to India*, where a white woman accuses her Indian host of sexually assaulting her in the Marabar Caves, despite her confusion as to whether the assault was 'real' or imagined. The history of travel is thus perpetuated in development literature and the genres of the Raj nostalgia novel and film, which contemporary development agencies recommend to workers as credible sources of information about Pakistan.

Up to this point in my history of transcultural interactions in the Northern Areas, I have concentrated on representations of indigenous people forged largely *in situ*. However, my research participants, especially the Britons, also 'know' these Others through domestic representations of expatriate Pakistanis.

Home Sites

Westerners have had a long hate-affair with Islam. Indeed, from its very inception, Islam has been represented as one of the most powerful threats to metropolitan civilizations. As the religion of the radically different 'East,' Christian and Jewish Europeans have usually dismissed Islam as an archaic set of superstitions practiced by fatalistic children tyrannized by their autocratic religious leaders into resisting European modernity (Al-Azmeh 1993; Bulbeck 1998; Kepel 1992; Said 1978, 1981). Consequently, Islam is stereotypically imagined by Westerners as composed of consenting masses of maniacal men, who keep their veiled women shut up at home, postured in prayer before a howling mullah. Anxiety about Islam has heightened with the increased migration of Muslims to Europe, North America, and Australia, the Iranian Revolution (1979), the Salman Rushdie affair (1989), the World Trade Center attack on September 11, 2001, and Western military incursions into Iraq and Afghanistan.

During a period of economic expansion in the 1950s and 1960s, the British government encouraged South Asian immigration to the United Kingdom (Al-Azmeh 1993; Ali 1992; Hassan 1995; Mohammad 1999). However, Muslim immigrants were structurally excluded from mainstream British life, and their cultures were artificially homogenized by official multicultural policies (Al-Azmeh 1993; Ali 1992; Mohammad 1999). These state forces prompted some immigrants to construct a cohesive Muslim 'community' and counteridentity to white non-Muslims as a way of consolidating their marginalized experiences. They employed what Aziz Al-Azmeh calls "hyperislamization" or an increased emphasis on religiosity to invent the notion of a distinctive "Muslim culture," which was then propagated within and claimed by the British Muslim community through many institutionalized practices. Visual boundaries of inclusion/exclusion were especially important to this invention. For example, modest styles of dress (the *shalwar kameez* and veil in particular), public exhibitions of piety and prayer, and the social detachment of British Muslims from mainstream modern life came to be signs of community membership. But cultural boundaries were also marked institutionally. Islamists called for gender-segregated schools, Muslim-only funfairs, Muslim medical advisory services, and an unelected Muslim parliament in order to conjure up the notion of an 'Islamic community' (Al-Azmeh 1993; Sahgal and Yuval-Davis 1992). So, the notion of a homogenous and essentially different 'Muslim culture' was invented simultaneously by white Britons through multiculturalist practices and by the Muslim diaspora's attempts at resistance.[19]

British Muslim women are central to the attempt to solidify a 'Muslim culture' and identity. Because women are regularly constructed as biological reproducers of the nation, carriers of culture, and bearers of the markers of group identity, their roles, activities, bodies, and sexualities are often controlled to serve the collective interest (see Cook 2001; Kandiyoti 1993; Yeganeh 1993; Yuval-Davis 1997; Yuval-Davis and Anthias 1989). Women's movements, behavior, and dress are used to signify differences between those who belong to the community and those who do not. In order to 'protect' Muslim women from corrupting 'Western' influences such as 'gender equality' and thus to preserve them as markers of the 'authentic' group, their communities compel them to remain primarily in the feminine space of home, attend schools for Pakistani Muslim girls, have arranged marriages with 'appropriate' Muslim men, marry young, be employed only in part-time work, and wear *shalwar kameez* and *hijab*.

My British research participants have experienced life among Pakistani expatriates in England; read European media representations of fanatical Islam, the Iranian revolution, the Rushdie affair, and the World Trade Center attack; noticed countless Muslim women wearing *shalwar kameez* and *hijab* on English streets; witnessed the social, sexual, and spatial exclusion of British Pakistani women from British life; and taught Pakistani children who belong to exclusive Muslim organizations and attend Muslim-only funfairs. These complex practices of domination and resistance are reduced to confirmations of the imperial narrative about the quintessential Otherness of Islam, Pakistanis, and Pakistan. However, the narrative becomes more 'real' to my participants as it is lived and experienced in Britain. British policies of multiculturalism, as well as feminist and antiracist discourses, have certainly shifted some white women's attitudes about Pakistani women in Britain, especially second-generation Muslim-English women. However, if women of Pakistani descent do not conform to European cultural norms (i.e., wear European clothes without veiling and have love marriages), then they usually continue to be viewed by most white women as the radically different Other. And the characteristics that appear to differentiate 'Muslim Pakistani women' and 'white Western women' serve to isolate them as two essentially different groups with nothing in common, with no point of contact, either at home or abroad.

Historic Citings

Each of these spheres of representation—colonial exploration literature, travel writing, tourist guidebooks, development manuals, and metropolitan constructs of Islam and expatriate Pakistanis—inform my research participants' perceptions of Gilgiti women. Western women have 'knowledge' of an irrational, oppressive Islam and of Muslim women who passively accept their subordination under Islam. These power-infused 'facts' prompt them to feel sorry for victimized local women, to understand them as essentially different from and subordinate to Western women, and to be grateful that all women do not experience the same fate. They effectively make patriarchal practices in North America, Europe, and Australia disappear from Western women's view.

Edward Said (1978) claims that much metropolitan knowledge of the Orient and its people, like the numerous representations I cite above, contributes to the phenomenon of Orientalism, a Western system of thought based on the notion of an indisputable difference between the Orient (Other) and the Occident (Self). Metropolitan authors effectively 'Other' the Orient when they methodically represent it as necessarily different from and subordinate to Europe. As we have seen, these representations are especially powerful due to the intrinsic citationality of Orientalism (see Brantlinger 1988; Christie 1994; Gregory 1995; Said 1978).[20] Said (1978, 40) explains that this authoritative Western knowledge about the Orient, which manifests itself as discourses of truth and power to dominate the Oriental Other, has historically been a companion piece to imperial power. The Orient is objectified through the "machinery of representation," and is thus transformed into a text that can be read, studied, known, judged, and managed by metropolitan audiences for particular metropolitan objectives, such as colonialism (Gregory 1995, 52).

Numerous authors have refined Said's work to demonstrate that the discursive formation of Orientalism is less monolithic, homogenizing, and totalizing than Said suggests (e.g., Breckenridge and van der Veer 1993; Jewitt 1995; Lowe 1991; Oddie 1994; Pinney 1989; Sprinker 1992).[21] These theorists argue that, rather than being cohesive and unified, Orientalism consists of interconnected, flexible, and heterogeneous discourses of power/knowledge that fracture and recombine in ways that alter them over time and space. The heterogeneity of Orientalism stems in part from Westerners' multiply constituted subjectivities. The contradictory nature of subject positionality enables some Westerners, women in particular, simultaneously to collaborate with and struggle against projects of colonialism, imperialism, and modernization in their writings and other social practices (Blunt 1994a, 1994b; McEwan 1994; Mills 1991, 1994; Pratt 1992; Robinson 1994). The heterogeneity of Orientalism also means that, while discourses of Otherness may have survived into the present, they have done so in contingent, ambivalent, and nonlinear ways.

The subsequent three chapters explore in detail this contradictory nature of subject positionality by tracing the multiple discourses, including Orientalism, which Western women exercise to constitute their subjectivities in Gilgit, with particular power effects. The bazaar story I narrated here about Western and indigenous women's lives in

Gilgit provides a context for this analysis; it evokes more fully the circumstances of data collection and analysis, situates me (as story writer) in the research setting and process, and foreshadows important perceptions, spaces, practices, people, institutions, and discursive histories that figure prominently in the narratives that follow.

Chapter Two

Vulnerable and Spatializing Subjects

Even for women who travel frequently, settling into life in an unfamiliar place abroad can be stressful. In Gilgit, Western women anxiously wonder, for example, how to dress, where to live, whom to trust, how to move about town, where to send their children to school, and what to do in their leisure time. In addition to contending with unexpected heat and cold, the lack of 'mod cons,' and the absence of other women in the streets, they have to learn new shopping procedures, including how to bargain for cheaper prices in a local language, walk deserted town streets at night when they can't get a lift, and be vigilant about contaminated food and water. Where can they find yeast, olives, and cheese? How should they interact with Gilgiti men at work? I recognized from my own experiences these women's narratives of feeling ill at ease in this unfamiliar social, cultural, and religious setting. Daily life in Gilgit can be overwhelming when you are never sure how indigenous people expect you to behave, but it is even worse when you are ambivalent about behaving as you think you should. These minute-to-minute uncertainties and stresses can accumulate to engender feelings of despair, alienation, and exhaustion, as well as tears of frustration. However, most of my research participants, who have large reserves of physical and emotional stamina, eventually learn to negotiate life in Gilgit in ways that allow them to minimize their stress, feel moderately relaxed, carve out spaces of comfort, and construct satisfactory lives and identities for themselves. In short, they devise strategies for 'stayin' alive.'

But many Western women—especially the young and single—remain edgy in Gilgit. This is not surprising as they are just 30 foreign women in a local community of 60,000, and they rarely encounter the roughly 40 percent of the population who are not men. Moreover, local languages and new social interaction rules are difficult to learn, and perceived Muslim vigilance over the separation of the sexes generates tensions that can be exhausting and annoying. But apart from and more intense than the types of social unease mentioned above is a haze of vulnerability, anxiety, and undetermined insecurity that permeates their lives. In most cases, women readily identified their social anxieties about interpersonal communications to me. However, the more generalized threat or vulnerability they experience is most often left only vaguely articulated. This palpable silence denotes an elusive sensitivity around which decisions are made and actions are taken.

If these women are anything like most of their counterparts in metropolitan centers, they also experience significant levels of insecurity at home. Research conducted in Australia, Britain, and the United States demonstrates that fear, especially of crime, is one of the most critical issues facing many women in these countries (Carcach and Mukherjee 1999; Koskela 1999; Madge 1997; Pain 1997, 2000; Smith 1987; Valentine 1989, 1991). Women often feel so vulnerable at home that they change their behaviors and spatial movements to prevent themselves and their family members from becoming victims of crime and violence; they avoid public space, stop traveling to work, disengage from social and sporting activities, and worry incessantly about their children's safety (American National Survey on Crime Prevention 2001; Carcach and Mukherjee 1999).[1] Many Western women have become what Ulrich Beck (1992) calls "risky subjects," social actors whose daily activities and sense of self are significantly informed by a fear of risks. When a sense of vulnerability to crime permeates women's lives and overshadows other concerns, it becomes intrinsic to everyday life. They spend much of their physical and psychic energies making themselves aware of risks and dealing with risk control. By enacting their perceptions and fears of risks in their everyday practices, women's sense of self and ways of understanding their relation to the world is constituted through risk.

Why do women in Britain, Australia, and the United States feel such high levels of vulnerability, risk, and fear? Because many women have a profound fear of sexual assault and perceive rape to be a risk that accompanies most other crimes, rape can be understood as a master risk that drives women's fear of violation (Dobbs 2000; Fisher and Sloan 2000). Women's fear of being violated is thus related to their fears of rape and of men. Although there is much feminist debate about which men women fear most, statistics from one Australian study show that middle-aged women who live with a male spouse fear for their personal safety more than young single women do (Carcach and Mukherjee 1999). But in Gilgit, older Western women do *not* feel more fearful than younger women. Married women, or those who have a male partner, usually feel *less* vulnerable than single women, although older women tend to feel safer than young women, even when they live alone. Western men are actually understood by Western women as protectors, as providing safe havens for them. Elena, a 29-year-old teacher trainer, describes the benefits of life in Gilgit with a fiancé this way:

> I usually jog from here to the eye hospital and back. I've done it alone quite a few times. I've never sort of experienced, well, once I remember three boys about 15, 16 that ran after me for about 100 yards, and then they all collapsed . . . It's nice now to have Peter along with me, because I suppose, you know, local guys would lurk and shout stuff before, but now I can just ignore them. Nobody even sort of *dares* to look if I'm with him.

Speaking about her spouse, Susan admits that "The balance has changed, I guess. Rick is more protective of me, and I would never allow that at the start, but it's easier here, and I don't want to fight all the time . . . Rick usually goes to the bazaar, just because it's easier. So if I can avoid *hassle* I will. It's just my survival instinct. It's one way of coping with the difficulties." And Andy, a single 26-year-old from Britain, thinks that

> the [Western] guys here are *incredible* . . . They just accept that it's easier for you to go to the bazaar with them, accept that it's easier to walk around with them . . . I thought I'd go back hating all men. And I thought that would be a struggle for me, to be able to return to England disliking men. But I think I won't, because the men I've met here, and I have to say now the *Western* men I've met here, have been so incredible and supportive that I think they have renewed my trust in many men.

These descriptions of Western men's protectiveness provoke several questions: From what are Western women being protected? To what difficulties and hassles are they alluding? Why do they distrust men who are not foreigners? Louise, a single 30-year-old, more directly addresses the source of her vulnerability: "[Returned volunteers] advised me to wear a wedding ring, so that [Gilgiti] men would think I was married, so I didn't get approached. That was good advice." Andy has a similar strategy: "I've often thought of lying and pretending that Karl was my boyfriend so that [Gilgiti] men would stop hassling me." Margaret, at 45, is the only woman who explicitly names the threat:

> Well you see, [younger women] are more threatened. They're afraid that the Pakistani men are going to take them off to bed. I feel the confidence

I have as a married woman, that they'll probably read me as far too old for it anyway, so I'm pretty *safe* . . . But I got a lift the other day with . . . a jeep full of men . . . I was a *little* bit concerned at one point, but, I'm probably being naive about being raped. I didn't really believe it would happen, but I'm sure it *could*.

Although there are important variations in women's perceptions of and responses to Gilgiti men due to their age and marital status that require analytical attention, many of my research participants, like most Western women at home, situate themselves as subjects in Gilgit in relation to matters of risk and vulnerability, particularly the fear of sexual harassment and abuse. To varying degrees they believe this risk diminishes if Gilgiti men think they are sexually claimed by a foreign male spouse. But while the *nature* of women's fear at home and abroad is similar, the *source* of threat in Gilgit has shifted. Indigenous men replace Western men as the primary source of women's sense of vulnerability.

More questions arise: Why do many Western women fear these men? Why do foreign men, who many Western women see as potential sexual harassers at home, become safe and trustworthy in Gilgit?[2] To answer these questions, I argue that women's vulnerable subjectivities and vaguely articulated moral panic are ambivalently forged in relation to perceived sexual dangers associated with (a) 'lascivious' racialized men, who pose an overriding threat to white women due to their sexual, cultural, and racial 'primitiveness,' and (b) local spaces, like the bazaar, that are associated with these men. As Ann Laura Stoler (1989, 636) argues for the colonial era,

While sexual fear may at base be a racial anxiety, we are still left to understand why it is through sexuality that such anxieties are expressed. If, as Sander Gilman claims, sexuality is the most salient marker of Otherness, organically representing racial difference, then we should not be surprised that colonial agents and colonized subjects express their contests—and vulnerabilities—in these terms.

This situation persists when contemporary Western women in post-colonial Gilgit fear 'primitive' male hypersexuality. I interpret women's experience of vulnerability as a discursive manifestation of their implicit fear of being ravaged by Gilgiti men, which permeates their subjectivities and sociospatial interactions with indigenous men. My research participants resist the threat of sexual violation, in part, by examining racialized men's behaviors for evidence of sexual excess and aberration. Through representational and surveillance practices, as well as spatial negotiations, Western women perpetuate—but also sometimes resist—eroticist and racist discourses about Other men that reinforce established social, sexual, and spatial boundaries. These boundaries, in turn, keep imperial hierarchies between indigenous men and Western women intact.[3]

I develop this argument by first reviewing how discourses of racialized sexuality developed in nineteenth-century colonial representations of the colonized. Following that I explain how Western women's sexual vulnerability is ambivalently shaped, how they react to 'dangerous' Gilgiti men with strategies of survival, and what discursive and boundary consequences these practices have. Moreover, I outline three spatial implications of the social and sexual boundaries formed in reaction to the 'lascivious male Other.' Social and sexual boundaries are realized spatially when (a) women's sexual fear is projected onto racialized spaces, as well as when (b) they attempt both to control spaces where dangerous Gilgiti men are absent, and (c) to create spaces where they feel safe. I conclude that, among Western women in Gilgit, sexually vulnerable subjectivities are spatially, as well as discursively constituted. Specifically, these 'vulnerable subjects' spatialize the social landscape to produce imperial fields of inclusion and exclusion.

Constructing the Danger

Many postcolonial theorists argue that the construct of the sexually dangerous male Other has had considerable currency among Europeans since the colonial era. Even earlier, in the Middle Ages, Europeans viewed racialized Others mainly as figures of their concupiscence (Gilman 1986; McClintock 1995; Pieterse 1992). However, not until the nineteenth century do we detect the deepening intersection of particular notions of race, sexuality, gender, and culture, which reified the construct (Levine 2003, 2004b; McClintock 1995; Young 1995). Dominant scientific and anthropological discourses about race and sexuality at this time merged and congealed particularly in scientific theories of hybridity. Many European scientific treatises were devoted to explicating theories either of monogenesis or polygenesis to determine how many human species existed. The final, but temporally elusive, test of these two competing theories was the continued fertility of the hybrids that resulted from the sexual union of two different 'races.' Hence, "theories of race in the nineteenth-century, by settling on the possibility or impossibility of hybridity, focused explicitly on the issue of sexuality and the issue of sexual unions between whites and blacks. Theories of race were thus also covert theories of desire" (Young 1995, 9).

These racial theories—founded largely on European fears of the sexualized Other and the notion that Europeans were morally, culturally, and sexually superior to other 'races'—were used as scientific 'proof' of a natural racial hierarchy. This hierarchy was established on the basis of degrees of racial civility, primarily an opposition between Europe's civilized moral/sexual order and a savage propensity toward sexual excess on the part of Others. As Mercer and Julien (1988, 107) argue, "sex is regarded as that thing which *par excellence* is a threat to the moral order of Western civilization. Hence, one is civilized at the expense of sexuality, and sexual at the expense of civilization. If the savage . . . is the absolute Other of civility then it must follow that he is endowed with the most monstrous and terrifying sexual proclivity." Colonialism, the civilizing mission *par excellence*, was thus rationalized as the route through which naked, uninhibited, impetuous 'savages' could be transformed into cultured individuals by Europeans whose sexuality was restrained by the manacle of Western civilization (see also Engels 1990; Levine 2003; Scully 1995). The 'white man's burden' of civilizing 'savages' through the spread of European culture consisted, in part, of taming the terrifying sexual proclivities of Others so as both to contain any threat to the Western moral order and to counteract cultural and sexual savagery abroad.

European attitudes toward 'noble and ignoble savages,' however, were not as straightforward as the civilizing mission might suggest. A profound ambivalence is embedded in the colonial fantasy of race and desire. Europeans were simultaneously attracted to the apparently extroverted sexuality of racialized Others and repelled by a 'primitiveness' threatening to the civilized world. As a result, racialized Others were both hypersexualized and marked as sexually taboo (Levine 2004a, 2004b; Sharpley-Whiting 1999; Stoler 1989; Sturma 2002; Young 1995). Frantz Fanon (1967) claims that the stereotype of the sexually devouring 'primitive' refers more to the ambivalent attitude of nineteenth-century Europeans toward their own sexuality than to actual experiences of Others. To allay their own anxieties, European fears and desires were projected, in exaggerated form, onto racialized Others (see also Bush 2004; Gilman 1986; McClintock 1995; Stoler 1995; Strobel 1991; Wiegman 1993). This ambivalent epistemic projection took many material forms, particularly in European women's interaction with colonized servants. In British India, for example, racialized men were understood as appealing and fascinating sources of domestic labor, but that appeal was complicated by apprehension when they transgressed the precarious boundaries of intimate social spaces such as women's hospital rooms and bedrooms. Even waiting for serving orders at the doors of these spaces often incited rape charges by European women (Stoler 1989).

According to the dominant discourse of racialized sexuality in this time period and imperial context, colonized men's unbounded sexuality followed from their immense penises and cultural 'primitivism' (McClintock 1995; Stoler 1997). Lore, often in the form of travel literature, had it that the sight of white women especially aroused colonized subjects' penises. Stoler (1997) notes that throughout many of the British colonies this sexual threat was labeled the 'Black Peril,' a fear that inspired initiatives such as ladies' riflery clubs and racialized rape laws. This threat, which had little correlation with incidences of rape (Bush 2004; McCulloch 2000), legitimated white control of colonized territory, surveillance of indigenous populations, and social, sexual, and spatial boundaries between colonizers and the colonized.

At the same time, racialized space was being sexualized in much travel and exploration literature. This aspect of sexual colonialism was enabled by the dominant Enlightenment epistemology of science, which depicted the accumulation of knowledge as a gendered power relation: male penetration into veiled female space/nature (Keller 1989, 1992; McClintock 1995). A masculinized European science emphasized power, control, and domination, which conjoined rhetorically the domination of nature with the image of nature as female. This conflation was already present in the writings of Francis Bacon. He writes, for example, that the science of his day can be represented as "leading you to Nature with all her children to bind her to your service and make her your slave" in a way that does not "merely exert a gentle guidance over nature's course; [scientists] have the power to conquer and subdue her, to shake her to her foundations" (as quoted in Keller 1989, 183). Science was coded in the sixteenth and seventeenth century to legitimate masculine knowers (mind) penetrating and disclosing female nature (earth, body, raw materials).

This thread of modern epistemology thus discursively constituted colonial travel and exploration as an "erotics of ravishment" (McClintock 1995, 22). Male travelers and explorers feminized the unknown and threatening boundaries they crossed. As men crossed borders into unfamiliar lands, they experienced an identity crisis, a sense of ambivalence fostered through the simultaneous "fantasy of conquest and dread of engulfment" (McClintock 1995, 27). The disavowed aspect of travelers' identity, their dread of engulfment, was projected onto the unknown territory, constructing both a feminized, 'virgin' space of conquest and a local population with the capacity to devour outsiders through their cannibalistic sexuality. Anne McClintock (1995, 22) calls this space the "porno-tropics," the European invention of racialized space as the "quintessential zone of sexual aberration and anomaly," and thus an ambivalent space of desire and fear.

In summary, hegemonic European discourses of racialized sexuality in the nineteenth century depicted Other men as lascivious 'primitives' with a dangerous appetite for white women, Other space as liminal places where sexual excess and deviance ran rampant, and white women as sexually vulnerable subjects in the porno-tropics. Through a series of conceptual ruptures and recuperations precipitated by multicultural, feminist, nationalist, anti-imperialist, human and civil rights, and anti-Muslim activities since the mid-nineteenth century, this cluster of attitudes and practices have been interwoven in inconsistent and discontinuous ways to both challenge constructs of racialized sexuality and reinscribe them in metropolitan centers. Notions of racialized sexuality have subsequently diversified, however ambivalently, in different places around the world, as race and sexuality articulate with gender and culture in different ways in different places to produce related, but distinct, manifestations.

Vulnerable Subjects in Gilgit

Jenny Sharpe (1991, 1993) and Nancy Paxton (1992) argue that official British reports of the 1857 Rebellion, India's largest anti-British uprising, and the many British novels written about it, solidified discourses of racialized sexuality in India. In these texts, authors who feared a political coup by the colonized depicted Indian men assaulting European women *en masse* during the insurgency. But Indian men were not the only group of male Others constructed as sexually dangerous in nineteenth-century European representations of the colonized. All racialized men, including, for example, Africans (Gilman 1986; Wiegman 1993), Arabs (Clancy-Smith 1998), and Fijians (Knapman 1986), could be expected to lust after white women. British colonial representations of Muslim Indian men, however, particularly emphasized this group's 'uncivilized' treatment of all women and, therefore, their distinct threat to colonial women (see, for example, Procida 2002; Ray 2002; Ware 1992, 142–146). Of all Indians, Muslim men were portrayed as the most terrible 'savages' who presented the greatest danger for and instigated the worst sexual crimes against white women, especially rape (Sharpe 1991). According to many British women colonists, Muslim men were rapists ipso facto (Sharpe 1993). This characterization was based, in part, on the misconception that the cloistering of women, which supposedly keeps women sexually unavailable and men sexually frustrated and eager, and polygamous marriage, a sign of men's exaggerated sexual need, were exclusively Muslim practices. Moreover, Muslim men's penchant for raping white women was thought to be rooted in their hypermasculinity. Particularly in the northwest frontier region of India, which was an important arena of British military struggle with Russia during the Great Game, Pathan Muslim men were, on the one hand, admired for their 'hypermasculine' bravery and war-like 'nature' by colonial administrators and explorers alike. On the other hand, these men were especially sexually suspect due to this very same 'savage' character.

Although the construct of racialized men's exaggerated sexuality is now more subtly experienced, invoked, and perpetuated through daily practice than it was in the nineteenth century, it remains a significant, if ambivalent, threat for Western women in Gilgit. Their vulnerability and fear are founded, in large part, on their racist representations of indigenous men as modern-day 'primitives.' According to their dominant discourse, Gilgiti men's primitiveness is the consequence of an amalgam of their religion, race, and spatial and cultural isolation in the high mountains of the Karakorum. Andy, for instance, accuses Gilgit men of participating in a "pathetic culture," where many of them are illiterate, intellectually inferior, lack a Western perspective, and force their women to practice antiquated Islamic social customs such as arranged and polygamous marriage and purdah. Many women claim that due to indigenous men's cultural backwardness, including their minimal contact with women who are not family members, and their sexually problematic social practices, they have not learned to interact in 'civilized' ways with Western women. Andy uses this example:

> That's one of the biggest differences between my Western friends . . . and my Pakistani male friends. My Pakistani male friends would take the keys away from me and unlock the house for me. I mean, can you imagine a Western man doing that?! It would just be *shocking* . . . He's been educated in Islamabad and all the rest of it, and he didn't think that women should be treated like servants. But it was just moving them to the Victorian stage[4] . . . I was quite shocked . . . Taking my keys, opening my house, hoping to get in, that's outrageous!

Her foreign male friends in Gilgit do not pose such cultural or sexual threats; she sees them as decent, trustworthy, and virtuous protectors.[5] Andy's trust in gallant Western men is shared by other women who rely on these men to provide safe passage to and from parties, run errands, and accompany them to the bazaar. I overheard my research participants ask their male counterparts to negotiate awkward work situations between Western women and Gilgiti men and comfortable spaces on appropriate transport. Women frequently claim that these men's "selfless" acts make it easier for them to travel more comfortably, eat at peace in restaurants, and enjoy more leisure activities without being ogled. As Rosemary states, "volunteers generally hang with the expat community. It's a 'known commodity' situation. If the single guy's around, you sort it out . . . these guys are very friendly, familiar, and you can dance with them, and it's *just not a problem*. It's very clear with these men."

Older and married women, however, sometimes see the civility of indigenous men differently. For example, Anguita, recalling the assistance she received from men in the bazaar after a fall, feels "spoiled for courtesy here. I think that Western men should take lessons of courtesy from Pakistani men. They are very *caring*, they're very *respectful* and polite . . . in the West women get treated very roughly. So, one has forgotten that this extreme courtesy exists. I like it."[6] Although Anguita barely tolerates the dirt in the bazaar, she seeks out daily interactions with shopkeepers who are considerate, have a good sense of humor, and fancy good bargaining challenges. I too have experienced most shopkeepers and male acquaintances and friends as exceptionally, and perhaps innocently, attentive, politely offering their hospitality and material support when it is appropriate, especially when I am traveling alone.

Although Evelyn also recognizes this solicitousness, she expresses her appreciation of Gilgiti men slightly differently. She conceded to me over supper that she is "glad that I don't know much about what happened with Bill Clinton and the cigar thing . . . I felt contaminated after knowing that about our *crud* culture . . . so I value the cultural purity here . . . I really admire it, you know, [Gilgiti people's] ability to be that pure, that simple, and to have such a good time doing things that are just *nice* things. So I value this exposure to purity." While this construct of the innocent and 'noble savage' is not unrelated to cultural primitivism, Evelyn represents it as sexually benign. Some younger women agree. For instance, Elena claims that Gilgiti men are sexually confused rather than aggressive: "[Western women] have the freedoms of a man here, but really we are still women, and we don't want to be *touched* by [Gilgiti men], like the way they touch their male friends or whatever. I think it's confusing for them." Instead of rejecting indigenous men as friends, Elena makes an effort to communicate her behavioral expectations to them. And once those expectations are laid out, she feels comfortable riding motorcycles and shopping with male workmates.

Anguita, Evelyn, and Elena may explicitly claim to be unthreatened by caring, polite, pure, simple, and confused Gilgiti men, disrupting the discourse of the 'lascivious male Other.' However, their statements imply some sexual anxiety that simultaneously recuperates this racist construct. By strategically feminizing a homogenized mass of Gilgiti men as 'pure' and 'nice,' Western women make themselves feel safer living among these men.

While a few of my research participants feel relatively at ease with Gilgiti men, the majority of Western women in Gilgit, to varying degrees, feel sexually and culturally threatened by them. Women connect what they interpret as cultural, particularly religious backwardness, to sexual backwardness. In an interview in 1999, René summarizes women's representations of indigenous people's cultural underdevelopment:

> They've basically lived in a *valley* all of their lives, and they haven't *seen* or experienced anything beyond their valley. Their experience is small, and their relationships are small. It means that even their imagination is small, and their ability to be creative seems to be small. They've been raised in a very small world, and, therefore, their perspective on life and their ability to dream is all small . . . They even don't have the TV and radio exposure that most people in the world do.

Amanda enlarges this view of a supposedly small, backward cultural landscape to include religious underdevelopment, and relates it to men's sexual deviance:

> And I always try to convince the girls to have a love marriage and not an arranged one, because they are so deceived afterwards . . . And then people told me how women never even take off their *kameez* [to have sex]. So then they're like a sex machine . . . Men should have those dolls, the rubber dolls with the hole here, and then they can use the doll and not goats or men or something else [for intercourse].[7] But they *really want* to get Westerners to have their sexual experiences before they get married to a virgin [Muslim woman] . . . Men's sexuality is so strong here.

She constructs 'sex-craved' local men as repressed yet yearning beings who long for liberating sexual experiences with foreign women. In so doing, Amanda recuperates the discourse of racialized sexuality, the notion that uncivilized male Others have an uncontrollable sexual appetite that is, if possible, directed at white women.

We also see in this quotation that the sexuality of Gilgiti men is constructed in relation to that of white women. As products of a generalized feminist politics in the West, my research participants understand themselves to be sexually liberated subjects who are free to make sexual choices without risking their respectability. However, they imagine that this same globally circulating discourse of Western women's sexual liberation is construed differently by Gilgiti men, as a sign of Western women's moral laxity, immodesty, and suspect character, especially in relation to dominant constructs of demure, spatially constrained, and sexually controlled Muslim women. My research participants frequently presume that Muslim men read their moral laxity and hypersexuality through their 'immodest' clothing (especially as portrayed in Hollywood movies), liberal access to divorce, greater choice in sexual partners, and independent presence in Pakistan without husbands or fathers to act as sexual chaperones. Consequently, Western women, in concert with this two-sided discourse of Western women's sexual liberation, are careful to dress appropriately in local clothing styles so as to forestall negative impacts on both their development work and daily interactions with local men.

In relating religious and sexual backwardness, some Western women avoid the Raja and Kashmiri Bazaars surrounding the Jangi mosque on Fridays as men finish their prayers. Instead of going to the nearby post office themselves on these days, they send office runners. They claim that, after hearing mullah "rantings," men leaving the mosque are even more likely to sexually harass women they meet in the street on their way home.

Rosemary similarly constructs Gilgiti men as questing for sex and enacting unnatural sexual practices: "The men talk about sex continually, even in my presence . . . I don't think they have a very good sex life, and I think that's responsible for a lot of their depressive tendencies. The men are *very* depressed, emotionally underdeveloped, and totally repressed. Nobody has a good *partnership*, there's nothing normal." When I asked for her opinion about the viability of relationships between Western women and Gilgiti men, she suggested they are doomed from the start: "You're never going to have the relationship you have with a Westerner, because sex is just a *physical* thing for [Gilgiti men]. It's not part of an emotional partnership, 90 percent of the time. So if you go out with a local, I don't think you'd be in for much fun anyway." Rosemary is careful to interact with indigenous men in abrupt and formal ways to indicate that she has no sexual interest in them.

Some of my research participants recognize that racialized men's supposedly perverse desires, especially for white women, are a consequence not only of their culture, race, and sexually repressive religion, but also of patriarchal Western media, which represent white women as sexually eager and available objects. Louise contends "When you put [men's dissatisfaction with their sexual relationships with their wives] with the media thing about Western women . . . then we're a good target for that sexual frustration." According to Julia, "These are men who never see any other women outside their close family. It's a society where sexuality is very repressed. And that there are particular notions about foreign women from the media, from films, from Baywatch, for example, absurdities like that, which we find a joke or insulting or meaningless. But that is what's there for them." And Susan asserts that "[Shouting at men] isn't going to stop them [leering]. They're watching Hindi movies, see Westerners on the television in bikinis. There's that stupid satellite beach party, dancing, all that sort of thing. And that's the images they see. That's what they think and want about Western women." Amanda is the only one of my research participants who sees the occasional film at the movie theater in Gilgit. Western women cannot understand the language of Bollywood and are usually uninterested in viewing 'sexy' James Bond and violent Rambo flicks. But more often they are concerned that sitting among Gilgiti men who are titillated by the scantily clad women in these films will provoke the audience to harass them in the theater.

While some of my research participants contrast what they see as infantile and sexually innocent or impotent Gilgiti men with themselves as sexually aware and liberated Western women, many others, such as Amanda, recuperate the racist myth of the Black Peril, the notion that 'underdeveloped' men have an uncontrollable sexual appetite that is heightened by white women. For example, many women's feelings of vulnerability are expressed in their impressions that all Gilgiti men have sexual designs on them, that they are the subject of all Gilgiti men's attentions and desires. I often reassured myself, while riding on public transport, that the men crowded around me in the back of the Suzuki van were more likely thinking about food prices, school tuition, family obligations, international politics, recreational activities, and work demands than about me. And René, despite being a married woman in her mid-thirties, supposes that Gilgiti shopkeepers have designs on her. She describes her interactions with them in the bazaar as sexually threatening:

> I feel uncomfortable I guess . . . I don't want them to get friendly with
> me. Maybe I'm on my guard or something, trying to keep them at a
> distance. Maybe I feel a little bit threatened or something. Like, you
> know, once they get friendly, then I start feeling like they're on the
> edge of abuse. You know, so I'm trying to protect myself from abusive
> behavior.

She thus restricts her interactions in the bazaar, shopping only at stores
run by men with a 'good reputation,' and avoiding friendly conversation and eye-to-eye contact even with these shopkeepers. Yet she mod-
erates this racism by claiming that Gilgiti men "are generally good and
well-intentioned and friendly and nice. I haven't really had any problems. I think it's my own personal attitude of just being a little cautious.
I don't really think it's *their* behavior so much."

Andy is similarly ambivalent. On the one hand, she depicts many of
her interactions with Gilgiti men as sexually harassing, even though
these same interactions seem 'natural' with Western men:

> When I first came here, and I started meeting some of these Pakistani
> guys, and they would *hug* me, you see . . . Because it feels *so* natural,
> you forget that they are crossing a boundary *too* big for them, that it's
> actually *huge* for them to even come in and rub your back. It's a big
> thing. Actually my *chowkidar* rubbed my back yesterday because I had
> a bad stomach . . . And I suddenly thought "That's actually not accept-
> able! What you're doing now is sexual harassment," because you've got
> to move it onto their terms. Because this is not Britain.

On the other hand, she describes Gilgiti men as "such kids" who are
"just *so* lovely that you can't find them threatening," "especially not
when you're teaching them." She insists that the men in her teaching
methodology classes, with whom she develops close working relationships, would never "do anything horrible" to her and that "I never feel
hassled, and there's nothing I don't feel I can cope with." However,
Andy experiences the fear of harassment by reacting to men who stare
at her: "[T]hat's what we were taught in self-defense at school. Make
yourself as *unsexually* attractive as possible. And then that gives you a
fighting chance. And so, yeah, I try *not* to make myself look sexual,
and I don't wear make-up . . . it's a thing about protecting yourself
and complying." She also avoids men's eyes and is careful not to laugh
or smile in public, fearing that this body language will be interpreted
by Gilgiti men as a sexual come-on.

As for how they represent indigenous men, Western women in Gilgit are conflicted. They experience the Black Peril, but simultaneously claim they are not threatened (or mistreated) by childlike or innocent indigenous men. Although there is little statistical evidence about sexual crime in Gilgit, not one of my research participants has reported a sexual harassment, assault, or rape case to the police or medical authorities. Indeed, as Louise claims, identifying anything beyond a 'pinch' as an instance of sexual harassment is often a difficult task for Western women in Gilgit. All of my research participants experience this ambivalence between fear and self-assurance, but some emphasize one pole or the other more strongly. This ambivalence can be interpreted as an intersection of imperial power and invincibility with gendered and racialized fears of sexual engulfment, projected onto Gilgiti men. Anguita describes this power as conceit: "I think conceitedness comes in, you know, in these gals. They're full of *themselves*. Westerners are full of themselves. They think they are the center of the world, always at the center of power." In some instances they are. For example, in Pakistan Jean uses what she calls "the 'memsahib approach to life,' and I really *hate* myself when I use it, like when I'm up against an administrative obstacle. They won't give me an airline ticket, and I'll start raising my voice . . . And it works, and they do it, and they break their own *rules* to do it . . . The privilege is undeserved, but it comes. And it's horrible, and it's colonial." Andy knows that "the trainees respect me because I'm white, not because I'm a woman. They wouldn't respect their own women in the same way. So my privilege comes with being a Westerner, always." And Marion recognizes the privilege of being a foreigner in the most mundane levels of everyday life: "I think that we [foreigners] get treated much better. When I go to the man who sells me fruits and vegetables, he picks me the best tomatoes. And I know he doesn't do it for locals. When you go to people's houses, they feed you, they give you everything, and I think that's because you're a foreigner." Most Western women, including myself, recognize that much of what they can do and is done for them in Gilgit is underwritten by global power and racial privilege. Their sense that they can change some official rules and assume respect from indigenous people conflicts with their feelings of vulnerability. They feel both powerful and fearful, both as women and white Westerners.

Western women's subject positions are also ambivalently constituted through conflicting senses of fear and desire. This is a racialized fear that articulates with a sexual fascination with Gilgiti men, as it did in the colonial context (Young, 1995). Although Andy distrusts Gilgiti men, she lets them rub her back and accompany her on long

walks in the mountains. Julia's conversations often revolved around
men's good looks and flattering attentions—the same men whose
stares in the streets she resents and who put her on edge in the bazaar.
Louise risks being alone with them: "I went to this one [Gilgiti] guy's
house once and . . . actually it was quite *unique* because I went there
and there was only a servant there. So I'm sure it was completely *inap-
propriate* for me to be there by myself [laughs]." She wants to gratify
her fascination with Gilgiti men, even though it is risky and at odds
with her understanding of local gendered norms of respectability.

Young women's ambivalence about the sexual proclivities of indigenous men seems to vary according to which men they know personally
and which they do not, although racism subtly infuses even their most
flattering representations. Women inspect men whom they frequently
see for evidence of sexual impropriety and excess. They note how and
where men sit on public transport, how they greet and interact with
white women, how they move around town, what spaces they occupy,
and how they attempt to be hospitable. Once shopkeepers and workmates prove they can "keep their hands to themselves," women know
them provisionally as safe and friendly, even 'lovely.' Elena, for example, says that "[Our tailors are] from Peshawar. Really nice. We've
known them for three years now, and they're really, really lovely boys.
They've measured us, and never touched us or *anything*. Just, you
know, appropriate all the time, never dodgy." Louise knows enough
men "that when I walk along someone will say hello to me at that
shop, and I *know* him. There's someone I could go to, and also there's
someone who knows who *I am* and who knows that I'm not just transient or walking through for the day. They're looking out for
me . . . It's the people I don't know that upset me here." Known men
are sometimes even seen as being like familial members who protect
Western women from unknown harassers. Julia describes a male
workmate who acquaints her with new shopkeepers and bazaars as
her *bhai* or 'brother.' And according to Jean, the principal of a local
school, "The male teachers are like my honorary brothers, and if we
travel on transport, they'll sit next to me as my male protectors.
They're a bit like family." Elena agrees: "I shared the house with one
of these [master trainers] when we were teaching a course together.
And he, you know, just like the others, he was friendly, lovely, we got
on brilliantly . . . The men that we work with, it's more like a family
where people look after you here." Even the older, married women feel
more at ease once they trust Gilgiti men not to be sexually aggressive.
Margaret, for instance, says that "There are shopkeepers who know
me and smile and say hello and invite me in for a cup of tea. It did

cross my mind to be wary when I was getting smiles from people I didn't know and I was responding exactly the way I would with shopkeepers. But it's *nice*, and it's nice to be recognized as a local and not a tourist." Margaret and Louise seem to believe that their relatively long-term presence in the area inspires feelings of familiarity, respect,[8] and responsibility for them in Gilgiti men, thus lessening the possibilities of sexual assault. Unknown and transient foreign women are apparently freer game.

For Western women living in Gilgit, being categorized as 'local' rather than 'tourist' is also important as their travels are motivated, in part, by a search for experiences with an 'authentic' premodern culture. They believe tourists, who are just passing through, cannot retrieve a more 'original' sense of self in the way they can, because tourists seldom have the opportunity to experience the 'authenticity' of local ways of life. Tourists, they contend, rarely experience indigenous lives that are ostensibly less materialistic, harried, and fragmented than their own. But becoming 'local' is not simply a matter of living in Gilgit for an extended period of time. To gain an intimate cultural viewpoint, Western women must be respected and trusted enough by some indigenous people to be granted admission into their lives. Many of my research participants crave respect from Gilgitis, and not only because they want to be accepted into local life. They believe respect is a resource Gilgiti men withhold from women, yet it is one that is owed to them due to their work expertise. Moreover, they think that if they are respected, Gilgiti men will more likely treat them as *bhajis*, elder sisters who deserve protection against unwanted sexual advances. Respect thus enables them to preserve some sense of control over their lives, as well as to experience local culture.

When I asked my research participants how they thought they could garner this respect in Gilgit, they agreed (with me and each other) that being married and having children were the most important steps in an Islamic setting where non-Muslim, Western women, especially those traveling unchaperoned, are morally suspect. They also suppose that respect is won by behaving appropriately, dressing properly, being in the right places, and not talking with or looking at strange men. But because development and missionary agencies equip Western women with only the most rudimentary knowledge of Gilgiti languages, religious groups, and social interaction and clothing etiquette, they must rely on hearsay and personal observations to judge what behaviors are (in)appropriate. Local guidance can be valuable, if offered, but miscommunications and misinterpretations abound. Western women are never sure how they should act and where they belong. I often experience negotiating daily life in Gilgit as if I were walking a tightrope without balancing aids. Is my *dupatta* placed properly? Is it appropriate for me to sit in the back of the Suzuki with Gilgiti men? Is my raucous laughter acceptable? Am I free to visit the library or see a movie without a chaperone? And my research participants are not even sure if they want to follow the rules they discern. But living with the nebulous consequences of noncompliance is frightening. Women are afraid Gilgiti people will disapprove of, laugh at, deride, and disrespect them for their social blunders. The fear of Gilgiti men goes along with fears of losing control and of transgression in a setting they do not fully comprehend.

Western women's lives in Gilgit are affected in many ways by their understandings of themselves as potential fools who may be 'out of place' and as prime targets for unknown men's sexual frustration. They restrict their daughters' interactions with indigenous men and boys, skirt places—like mosques, barber shops, and motorcycle garages—where Gilgiti men tend to congregate, avoid socializing with men outside work, and refuse to ride in crowded Suzukis, unless they can get a front seat in the ladies' compartment. But even that space is not necessarily safe or appropriate for white women. In Susan's opinion, "The [Suzuki] drivers are bloody cheeky . . . they touch you—there's the gear shift—and they have this *knack* of just sweeping [their hands] across your knees. And you can't actually ever say 'Oi, what are you doing?' because you're never sure whether it was an accident or not. I mean, I *know* that most of the time it's not. So I won't sit in the front."

Most notably, being sexually vulnerable and 'out of place' affects how women interpret the stares they get from men in public spaces. Once again, older women, like me, find the endless staring annoying, but they are seldom threatened by it. For instance, Dolly patronizingly admits that "We're different to these people, aren't we . . . There's so little in their lives, so I'm very tolerant of [the staring]." Evelyn is not bothered by it either. She says "I don't really mind. It doesn't bother me very much. I don't get very much of it, because I'm not often in the bazaar. I'm usually at work or in a vehicle. In a vehicle, they're staring at me because I'm driving, absolutely astounded that I'm driving, or staring at me because I seem to be talking to my husband or to another man very animatedly. You're not supposed to do that." Janneke asserts that "Although the men might stare a lot here in Gilgit, at least you know they're not going to really touch you. So most of the time I just wave and make them happy. [My husband] calls them my fan club." To Janet, Jean, and Fiona, men's "hostile," "disapproving," and "condescending" stares make the point that "good women" should never be found alone in the streets. But they do not find the stares sexually threatening. This talk about male stares in the bazaar, and the reasons for them, echo Annette Ackroyd's during the Raj:

> I do not think there were three women amongst the crowd, and certainly I was the only lady. In consequence of the infrequent appearance of a woman the people looked at me with profound amazement, and for the first time I realized how uncivilized are their notions about women. I read it in their eyes, not so much in the eyes of those who looked impertinently at me, for this is an expression not unknown to civilization! as in the blank wonder with which most scrutinized me. (as quoted in Ware 1996, 152)

While she is racially and culturally condescending, Ackroyd recognizes that women's experiences of sexual harassment are not restricted to the colonies.

Dolly, Evelyn, Janneke, Janet, Jean, and Fiona may not be threatened by men's stares, but Julia, who is single and 42, is quite ambivalent about what these stares mean and how she should deal with them:

> The hardest thing I have to deal with is being stared at by men all the time, you know, going to the bazaar and being maybe the only woman, or certainly one of the few women there. And just totally this being watched, being stared at. I haven't found any way of dealing with it effectively, and still it gets me angry sometimes. I really try to see that it's not me personally that they're looking at. It's some sort of [media] image.

Louise claims that "Often in the hotels it's just *staring*, and I can cope with that. It's in the bazaar, when they're looking in a *sexual way* that irks me." These differing characterizations of men's staring are summarized by 55-year-old Christine, who describes VSO volunteers being introduced to the Rawalpindi bazaar during their in-country training course:

> The men loved it, because they were free. But for us women, it was the first time that we'd really been *restricted*, and, although we could go out, it was a *nightmare*. Well, not so much for me, but for the younger ones, they were not at all happy with it. We got lots of attention and staring. And I think the attention probably bothers them more. I mean, I sort of feel that if they must stare, then they must. I think [younger women] see them as *lecherous* stares, a lot of them. And of course the cinema and television that they see here about the West is so slanted and selective. But I just thought it would mean just because I was fairer skinned and

blue eyed and had fairer hair than they, and they just found me a bit strange. I mean, I don't *like* it very much, but I think that's all it is.

Women's age and their sense of vulnerability are interconnected as, therefore, are their representations of male stares and attentions. Like Margaret, whom I cited above, Christine is relatively sure that, despite being fair-haired and skinned, she is "far too old" to be sexually attractive to indigenous men. But women's understanding of themselves as sexually undesirable to Gilgiti men does not contest the racist construct of the hypersexual or hyposexual/infantile male Other. Janet, at 56, confirms the dangers younger women face, but imagines she is safe:

> A lot of their problems feature on sex and sexual relationships, and how to deal with people who are *constantly* attracted to you . . . I mean it's just endless, endless, endless difficulties. I'm so thankful to be relieved of that . . . Although under my desk is the second of a series of gifts that a local policeman brings in. It started with a package of biscuits, and then it was a box of sweeties, and I think they are for [my colleague] and me. I certainly hope so. But then he came in yesterday, and this is a box of cherries from Skardu . . . Then I suddenly thought last night "Hummm. A box of sweeties, packet of biscuits, box of cherries. Should I be worried about this?" Maybe I'm making all sorts of terrible mistakes . . . I assume everyone thinks I'm a *granny* . . . So I'm guessing it's not an issue. But maybe I have something to learn. It might be dangerous.

Gilgiti men are possibly sexually interested, but older women have their 'granny status' as protection.

Western women who live with a sense of vulnerability in Gilgit spend much of their energies devising what they call "coping" or "survival" strategies to manage risk damage and control sexualized interactions and boundaries.[9] Their everyday risk management practices include never being alone with Gilgiti men and having very little interaction with them, taking children on shopping trips to the bazaar, and making it clear to men that they are married. Some women decide to walk and talk assertively, or ignore men's stares by wearing sunglasses and walkmans and by looking at the ground or the sky. Young women may form romantic relationships with Western men during their tenure in Gilgit, socialize only with expatriates, and present themselves as demure, including dressing modestly in *shalwar kameez* and *dupatta*, although few of them cover their heads. All women frequent familiar and 'safe' shops, which have been recommended by other Western women. And vehicles are not just convenient for trips to the bazaar in the heat or cold; they provide a physical barrier between Western women and Gilgiti men. Although Evelyn finds her deluxe 4X4 a "terrible insulator" when she drives through town, Andy prefers to shop out of the window of a vehicle, calling shopkeepers to the street to serve her, "because I feel contained, and safe, and in a place, and everyone understands white women in a vehicle." René, Janneke, and Abbie always negotiate shopping trips in their trucks, driving between shops and never wandering any length of the bazaar on foot. Lyn's boss lets her use a work jeep for shopping in hot weather: "The driver will take us up. Andy said it was because [our boss] was getting worried about how see-through our [summer] clothing was [laughs]. I don't know whether that's true or not."

Whether or not they enact their fears of the 'lascivious male Other' through their everyday practices, many Western women are ambivalently situated 'at risk' in Gilgit, which constitutes them as sexually vulnerable subjects and Muslim Gilgiti men as the source of sexual danger. Women who postulate the Black Peril try to avoid social interactions with Gilgiti men, and scrutinize their behaviors for signs of sexual excess and aberration. While self-surveillance of 'good' conduct is intense, the monitoring of sexually dangerous Others seems to be a more constant preoccupation. My research participants seem to be less threatened by indigenous men when they are able to exercise some control over social and sexualized interactions. Therefore, their sense of control is an important tool for ameliorating their fear of men's hypersexuality. But that control is quite elusive. Where it is less possible, such as in the bazaar or with unfamiliar men, the risk is seen as more acute, and hypersexuality and deviance is imagined to be greater. In these cases, the surveillance of racialized men allows women to gather 'proof' of misconduct, which can 'legitimate' social and sexual boundaries that keep Western women and Gilgiti men hierarchically separated.

Spatializing Subjects

These social and sexual boundaries also have three important spatial implications in Gilgit: the sexualization of racialized and gendered spaces; the control of spaces from which dangerous Gilgiti men are excluded; and the reterritorialization of group identity. First, women project their sexual fear of racialized men not only onto men's bodies, but also onto racialized spaces associated with these men, such as the bazaar. Women experience zones of sexual danger, those uncontrollable, enigmatic spaces that are occupied predominantly by the Gilgiti men whose sexuality they fear. Although Western women rely on vehicles and expatriate men to manage their interactions with indigenous men in the Gilgit bazaar, their inability to regulate most interactions there enhances their sense of the bazaar as a sexually threatening space. Boundaries between Western women and Gilgiti men are spatial, as well as social, making interpersonal and *public* boundaries of race and respectability necessary to enable women to watch over and control interactions (Stoler 1989). Gender, race, class, and sexual relations are thus negotiated through space in the contact zone.[10]

Public places in Gilgit that are predominantly male, like the bazaar, are experienced as threatening by Western women in a few spatially significant ways. First, because few Gilgiti women frequent the bazaar, many of my research participants are anxious about going there to shop, worrying that they are transgressing a gendered boundary and entering space off-limits to women. Western women often believe that this violation makes them appear to Gilgiti men as culturally insensitive, foolish, and 'loose' interlopers. They fear being disrespected even for having to buy groceries. Second, and relatedly, women experience the bazaar as a liminal porno-tropic because it is associated with ostensibly dangerous racialized men who disapprove of women visiting the bazaar. As a consequence of these two interrelated experiences of the bazaar, my research participants avoid it as much as possible.

Few women explicitly expressed fear of violence in the bazaar, but most of them indicated a vague sense of discomfort, some of it sexualized, fostered by this unmanageable racialized space. For instance, Elena prefers to avoid the "bit [of bazaar] past the Jama'at Khana and up towards the hospital. I don't like that place. It's all blokes hanging around there . . . It always seems like a bit of a seedy area . . . The blokes around there are really, they just stare at you and say stuff." Andy and Janet, attempting interactional control, are more comfortable calling shopkeepers to their vehicle in the street than venturing into shops and side alleys. Janneke goes

directly there and then back home. I don't go to the bazaar for the pleasure of walking in the streets. I'm not going to park the car at the Park Hotel and walk all the way . . . I don't go to the vegetable market by the Jama'at Khana. My cook will go there if he wants something, but I don't find it very pleasant, that dark little alley. It is cramped in there and they really stare at you. But for the rest of it, I seem to go almost anywhere, wherever I have to.

The enigmatic spaces, as well as faces, of the Gilgit bazaar can be sources of anxiety for Western women.

While older women are not entirely at ease in the bazaar, they are often more so than younger women. Marion, Amanda, Fiona, Christine, and I wander familiar sections of the bazaar alone and on foot. According to Christine, "The bazaar's a funny place. I mean, they're friendly enough, but I don't go. There's so little to buy. I certainly don't feel threatened there, it just isn't terribly friendly, mainly because they just don't notice you. Not that I *want* them to notice me, but you know what I mean." For two months I accompanied Amanda on her almost daily jaunts to the bazaar, where she explores the Gilgit library and drops by well-known shops to chat with male acquaintances over tea and to bargain for gem stones, rings, and cigarettes. But most younger women avoid the bazaar unless they are accompanied by friends, preferably foreign men. The favored strategy, if possible, is to send male spouses, friends, or servants for groceries and household supplies. Even some older women are anxious about being in the bazaar. Jane strays only short distances from her parked vehicle. Janet used to shop as little as possible, even when the cupboard was bare. But since a small shop opened beside her rented house, she can get the basic supplies she needs and be spared regular trips to the downtown bazaar. And René tries to stay away from the area altogether: "I avoid the bazaar at all costs. To me, going to the bazaar is something I hate. There's no fun in it at all. You know, you don't get to explore . . . So I go to the bazaar, in the car, and I come home as fast as I can . . . I'm ready to be defensive. I'm really very much on my guard." René's and Janneke's experience of the bazaar echoes that of Frances Shebbeare, a European woman who lived in late-nineteenth-century India:

> A rather alarming thing to do sometimes was to go down to the bazaar, to the Indian shops . . . I remember going a couple of times, I didn't like it at all. It was very uncomfortable. They had little tiny alleyways of streets, and everybody was crowding into them. They weren't going to hit you or steal, really, but it was just rather frightening. One hardly ever did it. If you wanted anything from the bazaar, you sent your bearer. (as quoted in Mills 1996, 13)

What these women experience in common is a guardedness and nebulous uneasiness about the bazaar. They claim not to be concerned about being robbed or beaten. Their anxiety seems to be generated by a sense that they are 'out of place' there, and by a vague sexual discomfort fostered by this inscrutable racialized space, with its dark, cramped, and unwelcoming recesses and faces.

All manner of allowances are made for Western women in the Gilgit bazaar, perhaps as a sign of respect, a way to increase custom from Westerners with money, or as an effect of racialized power relations. For example, they are not required to observe purdah there or to restrict themselves to ladies' shops. Shopkeepers are quick to order them tea and specialty supplies, as well as to arrange comfortable seating both in their shops and on passing Suzuki vans. Despite these allowances, visiting this gendered, racialized, and sexualized landscape takes some fortitude, because the spatial distance between 'us' and 'them' collapses. According to Robert Wilton (1998, 174), "Spatial separation facilitates the maintenance of social boundaries since it reifies perceived social differences between same and Other. Conversely, physical proximity challenges the legitimacy of social boundaries." By crossing into the porno-tropic, women relinquish some of their social, spatial, and cultural detachment from Gilgiti men and control over boundaries and interactions. Wynn Maggi (2001, 69) claims that "What makes boundaries so powerful is not that crossing them is unthinkable, but that it is so completely possible." This possibility of sexual, social, and cultural border crossings and boundary violations in the space of the bazaar informs women's identity as sexually vulnerable subjects, both spatially and socioculturally constituted. And women also spatially manage their fear of liminal local places by confining themselves to 'safe' sites and avoiding the bazaar as best they can.

Interestingly, some of my research participants describe relatively safe spaces within the larger Gilgit bazaar. Bazaars that are operated and frequented by Ismailis, such as the Jama'at Khana, Hunza, Cinema (or Nabi), and Airport Road Bazaars,[11] are preferred shopping sites for many women, not only because they work every day with Ismaili men and women at Aga Khan Foundation NGOs who encourage them to shop at their relatives' stores, but also because Western women see Ismaili men as less hypersexual than Sunni and Shia men.[12] Ismaili men seldom practice polygamous marriage or enforce purdah on their wives; their daughters are often as well educated as their sons; their wives are active in many sectors of the economy; and they have more interaction with women from outside their family circles. In short, Western women see these men as more sexually civilized.[13] While they too are racialized, my research participants believe that they respect women in ways that other local Muslim men do not. Due to their culturally enlightened attitudes, Ismaili shopkeepers and customers are understood as less sexually dangerous to foreign women. Hence, there is less reason to keep them at arm's length. It is often from these shopkeepers that women accept offers of tea, conversation, and assistance in the bazaar.

My research participants experience the bazaar as a sexually threatening place because it is an enigmatic space full of racialized men, and beyond Western women's control. But this space is not uniformly risky. It is constituted through a topography of threat: Ismaili bazaars are less threatening than Sunni and Shia bazaars, private transport is less intimidating than public vans, and the interior of shops and restaurants are less risky than open streets and dark alleys.

Second, the spatial implications of social and sexual boundaries between Gilgiti men and Western women are visible in women's attempts to manage spatially their fear of liminal local places, by creating and controlling 'safe' spaces where dangerous local men are absent, where they can relax and be themselves.[14] As Wilton (1998, 174) puts it, "physical proximity challenges the legitimacy of social [and sexual] boundaries." Western women have little freedom to control space in the Gilgit bazaar and to censure indigenous men's movements and behaviors there, although shopkeepers who need their business and coworkers who want their respect may acquiesce to women's limited attempts at control. Women react to their lack of control by partially and inconsistently withdrawing from spaces and types of interactions that seem especially threatening and unmanageable. They withdraw into enclaves, little pockets of safe space that are held back from a larger context of sexual insecurity and in which they can reestablish some sense of interactional control. These protected Western spaces of exclusion include their homes, the eye hospital swimming pool, and the Christian church. Women are at ease in these places, where their cultural identity and bourgeois morality is also confirmed. David Harvey (1996) and David Sibley (1995) argue that, while the spatialized control of supposedly threatening social groups is common in many societies, the power exercised to (re)assign people to particular places is usually left unacknowledged. To bypass this problem, it is helpful to conceptualize spaces as arenas of social conflict, where power organizes subjects in space and resistance involves attempts to forge alternative spatialities (Foucault 1977a; Harvey 1996; Pile 1997). Women are coping with sexual threat and elusive interactional control by creating 'alternative' local-free spaces, but their ability to exclude Gilgiti men from these spaces is enabled by racial and global power relations.

In the colonial era, racialized men's transgressions of social space incited rape charges by European women (Stoler 1989). While formal charges of this kind are rare today, foreign women often restrict Gilgiti men's access to these Western-only spaces because the latter are considered to be dangerously 'out of place' there. Indigenous men are not to enter Western private space unless they are deemed safe enough to be explicitly invited. However, as we have seen, Western women frequently violate boundaries between public and private space, even if it provokes danger. As their colonial counterparts did, they travel and work as 'honorary men,' and enter places, such as men's dining, work, and recreational areas, from which many indigenous women are normally excluded. My research participants, for example, commonly attend polo games, watch pool tournaments by the riverside, eat in nonpurdah sections of restaurants, and attend work conferences not restricted to women's educational issues. These spatial transgressions may invoke feelings of sexual insecurity in Western women at times, but they also instill a sense of superiority over spatially restricted Gilgiti women.

Women may feel compelled to distance themselves from Gilgiti men, even to create safe landscapes of exclusion, because they are often tense and nervous when these men are in Western women's spaces, whether that is in their homes, on a Suzuki van, or at an expatriate party or the pool. Lyn admits that "My coping strategy is to ignore men. Safe spaces are also important. Fortunately, except in crowds, it's not a problem. They don't get too close." Western women usually consider their homes to be their primary safe space, where Gilgiti men are not welcome. While Susan is comfortable, even enthusiastic about living with village families when she is on out-of-town teacher training courses, she is upset when indigenous men show up at her doorstep in Gilgit: "Rick's local friends will come in, and I'll be in shorts and a t-shirt, and I'll feel *really* uncomfortable in my own home . . . it doesn't matter when Rick says 'Don't worry about it. It doesn't make any difference.' That's easy for him to say. I mean, I can't *not* be bothered. I resent having to go into my bedroom and put my trousers on. I'm in my *own house*." Lyn, and apparently Elena, are similarly protective about their living spaces:

> [The VSO Urdu teacher] is staying at Andy's place. Elena and Peter are away, and the only reason he's staying there is because they *are* away, because Elena said "No, I don't *want* him here. I wear shorts and a singlet top around the house, and I don't care how *modern* he is, *I* would feel uncomfortable." And I was exactly the same. We have a spare room

there, and maybe it is OK. Maybe he wouldn't stare or whatever, but I just don't want him in my own home. I want to feel free to walk through the kitchen if I want to, looking like this, without worrying whether he's going to be offended or is going to stare or is going to do *whatever*.

Marion's concern about her physical proximity to Gilgiti men is apparent in her attempts to control space on public transport. She claims that "If the Suzuki pulls up and there's [men] in the front, I tell them to get into the back. And if they won't, I say fine, off they go, and I wait for another one. But last week I sat in the back . . . Someone talked to me, so I exchanged a few words only and then sort of stared into nothingness." Margaret manipulates space in the back seating compartment for the same reason, to create a safer environment for herself: "In a crowded Suzuki I generally use my bag or something to put between me and any man, so he doesn't touch me. And I look out the window and not catch anyone's eye."

As in colonial Rhodesia (see Tranberg Hansen 1992, 251–252), the safe space of a Western home in Gilgit is the usual site of expatriate parties, predominantly racially segregated affairs. Any excuse for a party will do. They occur frequently, and are eagerly anticipated by most foreigners. Birthday, anniversary, holiday, and *bon voyage* celebrations almost always entail dancing and drinking, and sometimes college party games. Women usually wear Western clothes to parties, little dresses, skirts, or pants and sleeveless tops that make them, as Jean says, "feel myself as a woman again, as a sexual being, you know, on my *own* terms." Young women are particularly fond of these occasions. Parties are held in safe places where these women can "let loose" and "be our Western selves," and where sexual segregation, with its incumbent anxieties, can also be checked at the door. Elena describes the atmosphere this way: "Parties are different. They're incredible, because it's like everyone just goes for it. The minute you come in the door you start dancing . . . We just dance all night. We just feel like 'Ah, what a release!'" The racially restricted space of the expatriate party releases women from local gender expectations and their surveillance activities, as well as the sexual anxieties induced by indigenous men and space.

For my research participants, parties are events that occur in spaces dissociated from work, shopping, and Gilgiti men, where they can shed their public roles and disregard the norms of Gilgit society. In expatriate homes they can dress and behave in ways that more closely approximate their ostensibly 'authentic' "Western selves." Here they relax the idealized, culturally correct self-portrait they present to

Gilgit society, and reveal aspects of themselves suppressed in those public performances due to local gender and religious norms. Home performances enable women to reinscribe a more comforting sense of themselves, to take a break from public performances, and to fortify themselves for subsequent 'contrived' self-characterizations. Women also experience parties as a release as they are freed from much of the sexual anxiety they feel when interacting with Gilgiti men and shopping in the bazaar. But older women, who experience these anxieties less frequently and intensely than young women, are more ambivalent about expatriate parties. Margaret admits that she is not "that excited about coming to parties. Sure, we get invited and we've come along, and we've enjoyed interacting with other VSOs. But we're not really here to spend our time with a whole lot of young volunteers from England, talking about the pub. I mean, that's not what we're here for. We want to be here so we can interact with the people."

But Margaret could mingle with some Gilgiti men at these parties. Volunteers usually invite a few known Ismaili men, their cultured and sexually trustworthy workmates. Julia describes parties she has attended in the past, when she worked with Ismaili men at AKRSP:

> A few local men were usually invited, coworkers invariably, people we'd known for some time and had some sort of trusting relationship with. And usually there was alcohol, smoking, dancing, mixed dancing . . . It was just very relaxing. Definitely something different and something like an escape from the surroundings . . . And there was quite an openness, a sense of behaving in a Western way in front of local men, and it was OK. They had had quite a lot of experience with working with Westerners. However, definitely there must be boundaries to that, and there are obviously boundaries in how you touch a member of the opposite sex, whether they were local or nonlocal, particularly in front of local men. Yeah. So I think, yeah, there still was some guardedness.

Women remain guarded for three main reasons. First, they work very closely with these men, and have come to respect their intelligence, work ethic, and commitment to improving the living conditions in their communities. As a result, Western women are anxious about offending Ismaili men by acting in ways that are denounced in that religious community. Second, women are concerned for themselves: they do not want Gilgiti men to think poorly of them. As I do, my research participants simply want all acquaintances to think well of them. But, as members of what they believe to be a superior culture and 'race,' as well as benefactors of Gilgiti people, this positive opinion is even more necessary. Moreover, they think disrespect will expose them to unwanted sexual advances. In this context of Western women's concerns for themselves and Ismaili men, the presence of Gilgiti men makes Western women ashamed of the way they dress, drink, and act at expatriate parties. Women internalize their interpretation of the indigenous male gaze. Although they seldom change the behaviors that help them recuperate a familiar sense of self, Western women experience the gaze by feeling guilty about and embarrassed by their party fun. The few women, like Amanda, Lyn, Dolly, and Marion, who seem to care very little about what Gilgiti men think of them, are more free of these constraints, and much less worried about sexual threat.

Third, my research participants are guarded about having Gilgiti men at parties, because they worry that even these culturally mature men, seeing them dancing with Western men, drinking alcohol, and wearing body-revealing clothing, might get "the wrong messages," might construe these party actions as signs of women's sexual eagerness, availability, and openness to men's sexual advances. Jean agrees with Julia that having Gilgiti men at parties is "[r]eally awkward. I'm pleased that they're there, because I don't like the concept of *angrezi*-only parties, but it is awkward, and I guess I don't mix so well, because I know I'm sending the wrong signals. It's like pretending I'm in Australia for the evening, stepping out. That's not so good either." For Christine, being exclusively with foreigners means "Just being able to *relax* and not always being slightly, you know, on edge, because of the culture." Having certain Gilgiti men at parties may ease some women's guilt about having racially restricted events, but it also means that parties are always somewhat disappointing for Western women. Not only is the music, environment, food and drink, clothing, and company not exactly right, but some Gilgitis are present to witness the aspects of self usually suppressed in public interactions. Public performances of self, which enact local, rather than metropolitan, norms of respectability and femininity, are thus compromised. Women worry that their party behavior will undermine any respect Gilgiti men have for them, which they believe can have sexually dangerous consequences.

As a sign of these worries, some women get extremely agitated when uninvited Gilgiti men attend expatriate parties. As described in chapter one, Andy is critical of Mustafa's relationship with his partner Uma. Andy sees him as a patriarchal and religious tyrant, a Sunni man who is sexually suspect due to his uncivilized treatment of his wife. Consequently, Andy likes to keep her distance from him: "Mustafa will come up and swim with us in the pool. But no, he should do his own thing. Or he'll go to Allan's party. I don't agree with it." Mustafa should literally know his place, which is safely away from foreigners. However, before I gained some understanding of how my research participants experience most Gilgiti men as 'out of place' in Western space, I invited Mustafa to accompany me to my first expatriate party. Andy was there too. She was so agitated about seeing him there that she left every room he entered so that she could avoid not only interacting with him, but being in the same space with him. She admitted feeling uncomfortable having him there while she was dancing and drinking in Western clothes, but did not consider how her social unease was imbricated in racism. This is one more example of how Western women maintain social, sexual, and spatial boundaries through their desire to restrict Gilgiti men's access to certain spaces, as they infuse indigenous men's bodies and spaces with their fears and organize their everyday practices around those risks.

The new cement swimming pool at the eye hospital, completed during the winter between my two field seasons, is another safe recreational space for Western women. It is hidden behind tall walls that merge into the existing structure of the expatriate public school. The pool, designed by the eye hospital doctor, is funded by Earth Mission and monthly user fees, and is off-limits to all Gilgiti people except the men who attend a school for the blind on hospital grounds. What is out of sight—white women in bathing suits—cannot offend Gilgiti sensibilities, embarrass Western women, or ignite sexual passions. Nevertheless, these indigenous men and Western women never use the pool at the same time. Swimming hours are allocated between 'us' and 'them' time. Women also have pool time separate from Western men, and adults separate from children. During summer family swimming hours, the pool is usually crowded with people from the four families who live in Gilgit semipermanently, as well as a few other foreigners who feel comfortable in this religious crowd and can afford the user fees.

After the children, women are the most enthusiastic pool members. Not only does the pool provide weather relief and child entertainment, it also allows more opportunities for what Evelyn calls "socializing within the expatriate community." Abbie likes "to have a lot more interaction with different people all the time. I don't want to see the same people constantly . . . The people who don't come to church, that aren't involved with that, that's why I come and enjoy the pool, because Allan is here, and you have been around, so I see some people here that wouldn't ever go to the service." But who constitutes "different people" is very limited: other foreigners who are not church goers. When I asked Evelyn about Gilgiti reactions to the pool, she naively said, without considering the labor that built it, "I don't know if they even know about it. How could they, really? It's behind a concrete wall."

Western women are enthusiastic about socializing at the pool for many reasons, but particularly because it is a space of racial exclusion, and thus of propriety and sexual safety. They agree with me that it takes some nerve and a significant shift in body image to replace a form-covering and honor-protecting *shalwar kameez* and *dupatta* with a revealing bathing suit at most swimming pools in Pakistan. However, they say it is easier at this pool, because there are no hotel staff to see them. I asked Evelyn "What difference does it make to you that Gilgiti men aren't coming here, in terms of putting on a bathing suit and relaxing around the pool?" She responded:

> Less than for other people, I think. We had a *lengthy* discussion about this the other night. And I think other women are more sensitive to those things than I am . . . Andy said something about Mustafa sitting around his pool and talking to women who are in small bathing suits. I never thought of it that way . . . I don't want to be a source of interest for people, but I'm also not hugely worried about it either. I'm somewhere in the middle.

It seems that women who go to the pool discuss this issue, and some are relieved to have a safe, local-free place to swim in conservative bathing suits or shorts and t-shirts without experiencing fear or provoking danger and disrespect. Others, like Evelyn, are less preoccupied with the risk of being seen. And yet other Western women still see the pool as unsafe, even though Gilgiti men are excluded, because they experience it as a site where normative Christian/bourgeois heterosexuality is subtly policed. Lyn, who keeps her distance from Christian folks and places and rejects their culturally and socially conservative attitudes toward gays and lesbians, speaks for herself, her partner Marion, and her close friend Dolly:

> That's probably one of the reasons I'm not too keen on going to the pool, because it's a churchy thing and all . . . In terms of that expat community, not a lot of heterosexual couples know what to do with lesbian couples. We were laughing with Dolly the other day about being oddities and perverts. She decided that she was considered the oddity and we were the perverts. People don't know . . . how to handle us, how to treat us, which is kind of amusing.

Racially exclusive spaces are still sexually threatening for some of my research participants, although the source of the threat and its consequences are different.

These vignettes about Suzuki rides, expatriate parties, and the exclusive swimming pool illustrate that once a power-constituted space is created, it acts as a site where subjectivities are reinscribed. Everyday interactions in space are social practices that reconstitute gendered, racialized, classed, and sexualized subjectivities by naturalizing the centrality or marginality of a particular subject (Ruddick 1996). Social practices spatialize to include some subjects but exclude Others from particular spaces. I think this can be understood, in the context of my research participants in Gilgit, through Julia Kristeva's (1982) treatment of the "abject," the psychically threatening not-Self that is incompletely separated from bodily ego in the unstable period of a child's identity formation and imperfectly excluded from consciousness. As it is not fully expelled, the abject lingers in the psyche to question and police the boundaries and desires of Self. Abjection is thus ambiguous: "while releasing a hold, it does not radically cut off the subject from what threatens it—on the contrary, abjection acknowledges it to be in perpetual danger" (Kristeva 1982, 9). The polluting dangers that threaten the boundaries of Self, which may include certain people, objects, and desires, are often cast out through social rituals of exclusion (Kristeva 1982, 65). The abject extends from corporeal boundaries to borders within social systems. Impure marginal subjects, who can potentially pollute central subjects or endanger them by crossing established social, cultural, and sexualized boundaries, threaten the psychic integrity of dominant subjects (see also Brah 1996; Butler 1993; Callaway 1987; Douglas 1966; Hoggett 1992; Wilton 1998). They are thus experienced as abject and socially excluded.

Jean is certain that Gilgiti men think Western women are immoral, and when I asked her how she reacts to this interpretation, she responded by recounting an experience of abjection that makes her physically shudder:

> Well, physical contact is interesting. I can't *stand* to touch a Pakistani man [shudders]. An *angrezi* man is fine. I don't even think about it. But if any Pakistani man touches me, I know it's not by accident, because I know they've got this invisible shield around them just as I do, and if they cross it, it's deliberately, even if it looks casual. But then again, I can switch it off with women. They lean into me, and I don't even notice that they're doing that.

For Jean, being physically close enough to touch a Pakistani man entails crossing a boundary that threatens her bodily, social, and psychic integrity. Andy has similar experiences: "Edward lived here, and he's a pretty close friend. If he's here I can open the door in my vest top, sleeveless top. And for Andrew, yeah I'd say for Andrew as well . . . But for the men around here, I need much more space between me and Pakistani men."

As Andy implies, abjection may also "be precipitated by challenges to the morphology of an existing spatiality" (Wilton 1998, 179). When dangerous subjects transgress spatial boundaries and challenge spatial orders, they also threaten borders of the Self, which can provoke feelings of anxiety and vulnerability in dominant subjects. On a trip to the bazaar, Jane was "getting meat, and our driver was with me. And where can I stand? There's a group of men around the butcher. You know, I can't get too close to them, because of my thing about, you know, some space around me. So, how do I stand? Do I put my rear into the aisle? I was trying to stand so, you know, that I wasn't vulnerable." Susan experiences Pakistani men as abject subjects who violate spatial boundaries while riding public transport: "When I [hurt myself] it was much easier to sit in the front [of the Suzuki]. But I *really* hated it. I *really* felt uncomfortable. I didn't want to be that close—this sounds terribly racist—I didn't want to be that close to a Pakistani man that I didn't know." Despite the countervailing influence of European antiracist discourses that problematize her feelings, Susan considers abject subjects, such as Mustafa and cramped Suzuki drivers, dangerously 'out of place' when they have transgressed established spatial boundaries. Chris Philo provides another example. He (1989, 259) argues that many suburban and gated communities in metropolitan settings are built as "closed spaces" that " 'purify' their surroundings by expelling and excluding [and thus reconstituting] all obviously different and 'polluting' categories of persons." That Western women resist being touched by Gilgiti men, are more comfortable with less social interaction with them, and are concerned to keep spatial boundaries between Self and Other intact suggests that my research participants frequently experience Gilgiti men as abject. Furthermore, they practice spatial exclusion when they feel threatened by a boundary loss between themselves and supposedly sexually dangerous Gilgiti men, and project that fear onto space, producing 'real' places with rigidly defined exclusionary boundaries.

These experiences of abjection sometimes prompt an unsettling psychic feedback response for my research participants. They experience Gilgiti men as abject subjects who are polluting and must be excluded, and, within the framework of that experience, identify certain interactions as evidence that these men find them equally abject. Men's ostensibly unfriendly and chastising stares, their incomprehensible verbal 'reprimands,' and their reluctance to eat food prepared by foreign women are construed as evidence that some Gilgiti men seek social and spatial distance from amoral, contaminating Western women. Janet, for instance, complains that Gilgiti men show "complete contempt" for her when they "cross to the opposite side of the street . . . It *completely* denies my existence on the street . . . Some days it does, it feels like hostility and contempt, just contempt for me as a person." And Rosemary, who lived with a local family during her first year in the Gilgit area, was allowed to eat lunch and dinner with the family as hers was a monotheist religion, but her host would never eat food she had prepared, not even tea. She suspected that because she was a non-Muslim woman, the things she touched were unclean to him. She then wondered if her big assigned private room in the house was not only a sensitive gesture, but also a way of providing social and spatial distance between clean and contaminated household members. These mutually constitutive experiences of abjection show how spatial boundaries may be constituted in relation to abject subjects. If marginal subjects are 'out of place' (not in their 'proper place'), they have transgressed social and spatial boundaries.

Third, social and sexual boundaries are realized spatially in women's attempts to reestablish their group identity. In order to cope in Gilgit's unfamiliar setting, Western women develop a situated group identity, in part, by constructing what Anne-Marie Fortier (1999, 41) calls "terrains of belonging." To reestablish a communal sense of identity and place, this displaced and fragmented group stitches together imagined commonalities—experiences of travel and alienation, a cultural history, an ethnicity, a beleaguered gender, a sexual ethos—which they project onto physical places like their Christian church, living spaces, and the swimming pool. Thus, they create a landscape of belonging, familiar places demarcated by exclusionary boundaries of the imagination that they rarely transgress in practice. The rituals enacted in these places cultivate a lived and embodied sense of belonging through repetitive practices.

Although it is not a place of belonging for all foreigners in Gilgit, the Christian church, raised every Saturday evening in a member's living room, is one exclusionary place where Western cultural rituals are enacted to restore a sense of the group's religious identity.[15] Church services include inspirational readings from American and British religious sources, homilies given by members on a rotational basis, and hymn singing to a piano accompaniment. Services are followed by pot-luck suppers, coffee, games, and conversation. As a testament to how the church activities instill in her a sense of belonging and identity in an unfamiliar setting, Fiona told me that

> The expat Christian community, there is a small group of us that get together once a week for some kind of church meeting and fellowship. They are a great source of comfort and company. We eat together, talk, and we've been away for weekends together . . . It's a very nice group of people here, but I think, because we're a small group, we have to support each other. Yeah, I certainly think that without that group I would have found it very difficult here.

René describes an important aspect of her social life in Gilgit: "Well, we always have church, which is a nice thing. We usually have some sort of pot-luck, so it's a social thing too . . . And, so people hang around, and eat dinner, and have a fun time just chatting and kids playing. So that's a nice social time in the week." Even people who are not church goers understand how practicing Christianity in church and on church outings fosters a comforting group identity. Marion discusses the Christian expatriates in Gilgit: "A lot of the other expats here are working at the eye hospital, which is a Christian organization, and their whole church services and everybody who is Christian goes to church . . . Their basic attitudes are similar, because they form a Christian group. They do things together, they go to church together, they have picnics together, etc." But the religious belonging cultivated in the church has exclusionary repercussions for other Westerners, as well as Muslim locals. Susan, in a slightly bitter tone, admits that

> The [nonvolunteer] Westerners don't really mix [with us] very much. There are a couple of reasons for it, I think. There are families, and there are non-families . . . And there's very much this expat-big salary/expat-volunteer salary. And there's also the church. So, many of the families that are here . . . are very involved in the church. I don't mind whether someone is Christian or not. It makes no difference to me. But that seems for them a very definite break.

Elena also feels excluded from many social activities within the Western community, because she is not a Christian: "I don't have much to do with those expats, really. I mean, it's always been a bit of a divide . . . They don't socialize with us. I was really shocked at first . . . They do their stuff together, because of their church, because they're all sort of church-going people, and that's how they like it."

These women are ambivalent both as foreigners who are 'in place' in the larger expatriate community, and who are excluded from this Western cultural space and identity. As Lise Nelson's (1999, 349) concept of "betweenness" and Pierre Bourdieu's (1990; Bourdieu and Wacquant 1992) notion of self-reflexivity suggest, the ambivalence generated as subjects move within and across fields of power and negotiate conflicting behavioral norms can prompt them to reflect on how they are situated within specific discourses. For example, Susan compares her Gilgiti, English-speaking neighbor to the Christian crowd: "She's very nice, and she's very accepting and always invites us to tea . . . She's nice and friendly to all of us, but most [Christian] people aren't. I don't know how different it would be if I was religious, if I was a practicing Christian . . . But what I've done is sort of make a life for myself without getting too attached to those other expats." When I asked her what she will remember most about her daily life in Gilgit, Susan replied "Being with the locals, more than anything. Not so much in Gilgit, but when you're out in the regions you're with local people all the time . . . People will do *anything* for you . . . And just how welcoming they are, how accepting they are of you and your strange ways, just welcoming you into their house, into their home . . . It's those people who will always really matter to me." By rejecting certain foreigners in favor of tolerant and caring locals, Susan may disrupt her sense of belonging among Westerners, her sense of self in a particular place. This questioning of 'us' versus 'them' sociospatial divisions may have some fracturing effect on Orientalist and racist power relations, potentially creating transformative 'spaces of hope.'

This example illustrates Nelson's (1999) theoretical claim that people can be discursively constituted *and* situated in discourses in ways that allow them to reflect consciously, yet partially, on the validity of those discursive norms, roles, and, ultimately, fields of power. By examining how they are situated in a network of discursive and social fields, people can engage with the multiple meanings of their past and present practices. This situated, shifting, and partial knowledge enables them to reflect on and become aware, to some extent, of how and why they perform their subjectivities as they do. And in doing so, they may identify new avenues for actively intervening in the process of self-constitution. Rather than reinscribing a fully conscious subject, as Butler (1993) and Foucault (1977a, 1978, 1980) fear, "nuanced conceptualizations of 'self-reflexivity' . . . open up discussions of how to navigate between the constitutive power of discourse and the historical and geographical embeddedness of thinking subjects" (Nelson 1999, 350).

Western cultural identity is also reterritorialized in expatriate homes. Parties are held here not only because they are safe places away from Gilgiti men, but also because they are terrains of belonging, sites of the construction of cultural authenticity. Salaried development workers usually spend considerable amounts of money remaking local rented houses into Western homes. As I sat in their kitchens drinking coffee, I noticed painted walls, tile backsplashes, wall paper borders, imported curtains, Western appliances and furniture, and even the occasional chandelier and air-conditioning unit. Most freezers I glanced into were full of fish, hotdogs, feta cheese, and stewing beef, all packaged and purchased in Islamabad. In this type of physical environment, expatriates can feel at ease in Western clothes and let their hair down at a party. Parties I attended usually began with a pasta dinner, complete with cheese and olives. But once someone turned on the music, or the foreign men who formed a rock and roll band began to play, the food was replaced with alcohol[16] and lively mixed dancing. The combination of food, drink, clothing, decor, and music allows Western women to "be ourselves," meaning that in the racially restricted place of home they practice rituals that instill a sense of belonging and cultural identity. These spatially situated practices of cultural identity, like those enacted in the church, illustrate "how bodies constitute the raw material through which place is both experienced and inscribed, and how bodies, in turn, are inscribed by and become the signifiers of particular notions of place" (Fortier 1999, 48). Bodies and spaces are mutually constitutive in efforts to reimagine a comforting, if exclusionary, cultural identity.

A Vulnerable Imperial Logic?

Many Western women are ambivalently constituted as sexually vulnerable subjects in postcolonial Gilgit through the confluence of discourses of race, sexuality, and culture. These articulated realms of experience constitute their subjectivities as women and represent indigenous men as cultural and sexual primitives, while they help women manage the sexual dangers posed by 'lascivious' racialized men and negotiate local spaces associated with them. Thus, although there are some significant variations based on differences of age and marital status, women's subjectivities are often forged in relation to a discourse of sexually dangerous Gilgiti men, even when some women use their age and relationships with Western men to mitigate the perception of an omnipresent risk. Women fear male hypersexuality and resist that threat by scrutinizing racialized men for evidence of sexual danger or safety, and by controlling social interactions and ordering space to exclude threatening and—I would argue—abject Others. These risk-infused strategies concentrate women's fears in Gilgiti men's bodies with oppressive imperial effects.

As unintended effects of some women's practices, the discursive constructs of 'lascivious racialized men' and 'sexually vulnerable white woman' are perpetuated, along with social, sexual, and spatial boundaries between Western women and indigenous men that have been unevenly preserved and contested since the colonial era. But do women's practices of surveillance and spatial control have a significant impact on social relations in Gilgit? Will oglers in the bazaar and Suzuki passengers change their behaviors and spatial movements to accommodate Western women? Do Gilgiti men care what foreign women want and expect from them in a context where women are often socially and spatially restricted? Only when my research participants have regular encounters with familiar Gilgiti men (i.e., shopkeepers and workmates) can they censure and discipline men's behaviors. Women exercise power to exclude Gilgiti men from certain spaces like the pool and parties and their homes, but that is a restricted form of control. However, it demonstrates that an imperial logic, which often has large-scale exclusionary social effects in a globalized world, continues to operate in this postcolonial space alongside shifting discourses of power that have been inconsistently circulated by Westerners over time.

But that recuperated imperial logic and power, with the boundaries they sustain, are destabilized, perhaps even ruptured at times, by those of my research participants who are marginalized within the expatriate community by their sexual identity as lesbians, or excluded from particular sites of Western identity production, such as the pool and the Christian church. Sometimes they disrupt discourses and practices of race and culture by aligning themselves more closely with locals than with foreigners, especially when they appreciate inclusive Gilgiti hospitality, and reject the cultural identity reproduced by other Westerners. Moreover, once my research participants represent particular Gilgiti men and specific local spaces as safe and "lovely," they contest discourses of sexual vulnerability and the generalizability of women's fear. After they have spent time with Ismaili men who 'prove' themselves trustworthy and safe, women often develop delicate feelings toward them, and, therefore, may be less likely to assume *all* indigenous men will be sexually dangerous. And once the threat diminishes, however slightly, boundaries become less rigid and border crossings are easier. These behavioral shifts may be understood, drawing on Certeau (1984), as tactics that emerge from the margins, practices that occur at the limits of power and can emerge creatively from the margins to challenge dominant social relations. Rather than always denying difference, some women are able to accommodate it, even adapt themselves to it. They transform some of their behaviors and attitudes into nonreactive practices, such as affiliating and sympathizing with locals and not with foreigners, that temper the experience of sexual and spatial threat, abjection, and cultural alienation, and thus partially disrupt dominant discourses of power.

Chapter Three

'Free' Travelers and Developers Navigating Boundaries

That the native does not like the tourist is not hard to explain. For every native of every place is a potential tourist, and every tourist is a native of somewhere. Every native everywhere lives a life of overwhelming and crushing banality and boredom and desperation and depression, and every deed, good or bad, is an attempt to forget this. Every native would like to find a way out, every native would like a rest, every native would like a tour. But some natives—most natives in the world—cannot go anywhere. They are too poor. They are too poor to go anywhere. They are too poor to escape the reality of their lives, and they are too poor to live properly in the place where they live, which is the very place you, the tourist, want to go—so when the natives see you, the tourist, they envy you, they envy your ability to leave your own banality and boredom, they envy your ability to turn their own banality and boredom into a source of pleasure for yourself. (Kincaid 1988, 18–19)

Why Do They Come to Gilgit?

When the Karakorum Highway opened in 1978, the majestic landscapes of Pakistan's Northern Areas, especially in those valleys adjacent to the road, became more accessible to travelers. Previously, relatively few Western or lowland Pakistani travelers made their way north along the ancient trading routes through the mountains. Serious mountain travel was undertaken mainly by explorers, climbers, and adventurers willing to arrange lengthy and physically demanding expeditions. But the paved road has made traveling less cumbersome, making the Gilgit area a new site of development activity, as well as an increasingly popular tourist destination for both national and international travelers interested in mountaineering, trekking, skiing, sightseeing, fishing, and cultural tourism (Development Research Group 1995). Despite Western Islamophobia and Pakistan's political and economic instabilities, sectarian violence, wars with India, and nuclear testings, there has been a fairly consistent increase in the number of foreign and expatriate national tourists to Pakistan since 1983 (World Tourism Organization 2001). Approximately 7.4 percent of Pakistan's foreign visitors go to the Northern Areas, and in the early 1990s that number reached between 25,000 and 30,000 tourists (Butz 2002; Friedberger and Pooley 1995; Mock and O'Neil 1996b). Almost all travelers interested in seeing legendary Hunza—the famed Shangrila— and Rakaposhi (7788m), Ultar (7388m), Diran (7257m) and Golden Peak (7027m) mountains, or in getting to China over the Kunjerab Pass need to proceed through Gilgit.

Most foreign tourists to the Northern Areas are 'pulled' there: they are drawn by the beauty of the landscape, mountain activities such as trekking and climbing, adventure travel, and the prospect of observing 'premodern' mountain people and cultures, all at a reasonable cost. Tourists on agency-organized trekking trips, cultural tours, and jeep safaris usually spend less than a month in Pakistan, but climbers and independent low-budget backpackers can be in the country for months. Even though some of these foreigners are on lengthy holidays in the Northern Areas, their experiences with indigenous people and cultures are often mocked by my research participants. In Louise's words, "As a tourist you just pop in and out, but I want to do and see a little bit more than *that*." Like most Westerners living in Gilgit, my research participants think their extended residence in the area, as well as their working and personal relationships with indigenous people, give them a privileged insight into local life that a tourist can never attain. Foreign development workers, who are offended to be called 'tourists' or 'travelers,'[1] are convinced they have a more accurate understanding of 'authentic' Gilgiti society and lifeworlds than do these other visitors. During her four years in Gilgit, Elena has learned that "the way people are here is so different than the way the people I know at home are. It's just a whole new race of human beings. They show what [Western] people *could* be like." In contrast to tourists and travelers, development workers situate themselves as part of the social landscape, as quasi-locals who observe from the 'inside' out, rather than from the 'outside' in. In their aesthetic sensibilities and "cultural discretion,"[2] my research participants align themselves with indigenous people and against other tourists for purposes of self-constitution and differentiation. They distance themselves from other Western travelers, because these foreigners threaten their efforts to cultivate unique and independent selves.

Many Western women in Gilgit also distance themselves from tourists by stressing their own travels to the Northern Areas; they are here not simply to holiday but to help, to work for a Western-funded NGO with the goal of improving local living conditions. This "politics of virtue" (Mindry 2001, 1193) adds another 'pull' toward travel abroad. It is the 'official' reason most of my research participants give for why they decided to come to Gilgit, even against their better judgment. And it is a noble reason. Most women told me they feel lucky to have had a good education, comparatively satisfying and well-paid work, and materially abundant lives in the West. They want to share that *kismat* with less fortunate people by putting their expertise to work in a developing country. Elena told me that

> I've been really lucky, because I've done anything I wanted to do . . . I wanted to go to university . . . and become a teacher, and I was able to do that. And then, I always wanted to live and work abroad somewhere, but not earn like loads of money . . . I feel, honestly, that we're still much better off than a lot of people here, but a local wage is what I wanted. So I'd describe myself as being very lucky. And I want to give something back.

Joan has also wanted to help for a long time: "I think there were *lots* of reasons why I got to the point of deciding to come. I'd always had it in my mind that I would like to come and work in a developing country— that's not Pakistan, just a developing country—at some point in my career, to help." Margaret has come to Pakistan as a volunteer development worker to repay her debt to people in other developing countries who have been overwhelmingly generous: "I've done a lot of traveling [in developing countries], and I'd reached the stage where I said 'I've got so much out of my experiences, how can I give back?' And this seemed to be a way." Fiona supposes that her "motivation was that I was looking for what God wanted me to do to help and where to do it. I'd been in Zimbabwe before this, and I came here first to help out for a couple of weeks. And they were looking for [someone] to run the organization, so it seemed to be where I should be."

Most women development workers in Gilgit who come through VSO would have preferred postings in Africa, China, or South East Asia, but they came to Pakistan, sometimes reluctantly, because of the attractiveness of the job description. When I asked Lyn how she landed in Gilgit, she replied

> Bad luck I suppose [laughs]. [Marion and I] wanted to go back to Africa, but a job fell through at the last minute. And then VSO offered us a position in the Northern Areas, and we said 'No way!' There was no way we'd work in Pakistan. But they sent us the job description anyway,

which was a downfall, because the job description was so good. It was exactly what we wanted . . . So we thought "Oh, this will be a change. We'll give it a go."

Marion is more specific about the down-side of coming to Pakistan:

If you'd asked us where we were going, we would have said that Pakistan was the last place *on earth*. There was nothing about it to appeal and all sorts of things to put us off, like it was a Muslim society, not just religion, but a Muslim state-sponsored society . . . We had a priority to work with women's and girls' education, and this job met that criteria and sounded interesting enough, even though it wasn't in Africa. And it was either this or thinking of something else to do.

Julia remembers "thinking about being a VSO when I was very young. It had been in the back of my mind for a long time . . . When it got closer to an actual placement they said 'Where would you like to go?' And I said 'I'd like to go to a non-Muslim Asian country' . . . There's a lot of prejudice about Pakistan in the West, and definitely it wasn't somewhere that I wanted to come to." Like most Western women in Gilgit, Lyn, Marion, and Julia wanted to do good works, but were concerned about how difficult that would be as foreign women in a Muslim society, where women are reputedly not well treated or respected.

This philanthropic pull factor—for privileged Western women to provide for the needy—can be understood as a positive aspect of a discourse of bourgeois morality and femininity, whether the benefactors are inspired to do the devout work of development or Christianity. However, ironically and in keeping with colonial-era discourses and practices, Western women who come to improve Gilgiti women's education and health and to transform sociocultural systems are often uninterested in the everyday lives of Gilgiti Muslim women—although they claim to know what those lives are like (Burton 1990; Chaudhuri 2002; Melman 1992; Ramusack 1990; Ware 1992).[3] Abbie is "amazed that I can even live here, because it just doesn't interest me hardly at all. I really like [the men I work with] . . . And I've liked a lot of the expatriate people. But I don't feel any draw to the local women. I don't want to go and visit my neighbors. I don't do any of that kind of stuff. I probably won't. I feel like I have to have my life separate from them." For many of my research participants, Muslim women constitute 'the needy,' as opposed to poor, yet interesting, people with compelling lives. They often serve as foils against whom Western women, through "transnational moralizing discourses" (Mindry 2001, 1189), imagine themselves as self-sacrificing benefactors.

As Joan suggests above, Western women usually have multiple reasons for traveling to Gilgit. Altruism is often mixed with "selfishness," as Helen Callaway (1987, 41) suggests, which makes Evelyn

> realize that nobody has 100 percent purity in their life. I was interested to meet a Canadian woman who graduated from Development Studies at Trent [University] . . . She's in Afghanistan now, so she'll get it out of her system and then go home. Even there, is she achieving purity, are her motives completely pure? I have no idea. So I'm at the point where I can't define who's really doing something because they want to help people, or, as in my case, a big part of the reason for me taking the job was that I needed the *money*. It's just *disgusting* to me [laughs]. So I would probably work for a development agency again, because I think I want to do good work. But I can only take that so far.

'Push' factors, such as a lack of employment opportunities in Canada, thus constitute another context of travel for Western women,[4] one that can be embarrassing for my research participants to acknowledge.[5] For example, when I asked Susan why she decided to do volunteer work she said "It sounds a bit *crass*. I wanted to do something, I wanted to feel worthwhile . . . If I'm honest, I'd say the main reason I'm here is for selfish reasons. The job opportunity is *great*. We get a lot of professional freedom here, which you don't get in Britain." Dolly also "needed to do something that's *worthwhile*. That sounds really *pious*, doesn't it . . . That's for me, never mind anyone that's on the receiving end. That's trying to be honest about it. It actually fulfills something in me. I only feel half a person if that's not happening." Louise feels the need to concede "the selfish reasons [for coming]. There are some. I'm not here just for the good of the world. I'm here for the good of me, too, to find out about a culture, and actually learn about it and live it, and feel that perhaps I can put something back in, as well as take." Jean suggests that the altruistic aspect of her work is even downplayed in favor of personal development by the volunteer agency that sent her:

> Australian Volunteers International has this other philosophy that, because you're bringing home the cultural experience, it's fine if you live in a community in a fairly harmless way. It doesn't really matter if my skills are transferred. That's the way it's working out in my placement . . . I'm having a good cultural experience . . . Ideally AVI would like both to happen, the cultural exchange and the skills transfer, but, in most cases, just the one happens.

Although 'push' and 'pull' factors intermingle in every woman's life, my research participants tended in our interviews to stress what 'pushed' them to Gilgit. Their remarks hint at some of the overlapping events, situations, discourses, and 'selfish' desires that forced them away from home. Janneke, Abbie, and René followed their spouses, who were traveling on development work tours. René and her partner, like Evelyn and Amanda, were also unable to find well-paid jobs at home. In Evelyn's words, they came "chasing money." But this job offered Evelyn more than an income: "I'm a bigger fish in this small pond than I was at home. I found that my career was nose-diving in Canada . . . Discretionary money and consultancies were drying up. It was impossible [to find that work] anymore. So *here*, I revived my career." Other women, such as Janet, Rosemary, and Rose, were fleeing an empty nest. Troubling family relations, including divorces, elderly parent nursing responsibilities, and bereavements, compelled Joan and Christine to leave home, and reinforced Rose's desire to migrate.

When reflecting on her reasons for traveling, Christine thinks she has

> lived at home, in England, much more on the *surface*, and I was very good at off-loading things onto other people . . . Being here has *cleared* my mind. I needed to get away from home anyway, for all sorts of emotional baggage reasons, and living here, on my own, has been very good, because I've been forced to actually think about things, rather than forget about it or off-load onto somebody else . . . Whereas here, I've sorted out day-to-day problems, but I've also sorted *myself* out . . . It's made me much more, what, self-reliant. Is that the word? Certainly more self-confident, because you realize you can *deal* with things, whereas before I always felt that I actually had to lean on somebody else to deal with something. Now I know that I *can* do it.

Becoming more self-reliant and assured fulfills one of my participants' common 'emancipation aspirations,' mentioned in most of their departure stories.[6] These aspirations, which women attempt to fulfill through travel, include an urge to escape dull lives,[7] a quest for self-reliance, and a desire to prove themselves strong, independent, and liberated. Janet concludes that her experiences in Gilgit have improved her preparedness for old age:

> I think that anything you go through that involves some degree of hardship only builds confidence really. And I think that is a thing that, as I'm conscious of going into old age, is a really good thing. If I've had an experience of having just *physically* survived, it brings me into a slightly more confident old age . . . I had hoped that as an outcome that I would

be a *stronger* and less *fearful* old lady, going into old age knowing that you can cope with things and not be fearful . . . And the knowledge that I can live *very* simply, contentedly is a very powerful thing.

Gaining confidence through travel has helped Dolly escape an increasingly dull life at home: "I don't think I can settle back in the UK. It's just not enough . . . I think I need the variety . . . I don't know what's missing there. I've outgrown those experiences . . . I wouldn't be satisfied to go back to that. But I'd quite like to go and see what's around a few more corners, now that I've got more experience of traveling and more confidence about traveling with VSO." Rosemary's desire for a life change pushed her into travel. She thinks that "probably I was looking for a total change. I raised my kids—they're independent—and I'd had enough of my job . . . It was quite the life, and I just didn't need to be there anymore. So I was very open for something different, and this place is affordable on what I've got . . . And I figured it would be a good way to live life, teach and be here. And then I could travel in the off season, through Asia." Christine's search for self-reliance is a reflection of her need for independence: "I don't think I'd ever really looked at myself until I decided I wanted a change . . . I think I'd been really busy from marriage onwards, keeping all the balls in the air, keeping the plates spinning, because it's silly with three kids and a job and a husband and a large house and everything. There *really* wasn't much time for *me*. That's probably why I needed to break out of the rut." Joan was also hoping to realize some personal freedom through travel:

> I got heavily involved with my career, which became very important to me. And then, about 17 years ago now, I had to become the main caregiver to my mother who was disabled . . . She died in 1997. And I *suddenly* thought that if I just carry on doing what I had been doing—I was principal of a large school in London—I'm not using the freedom that I have felt deprived of over these years. So it was *that* thought that got me here.

Traveling seems not to be a solely altruistic venture for Western women in Gilgit, but also, as it was in the colonial era, a means for white women dissatisfied with their lives to achieve some sense of personal freedom and to constitute themselves as full, independent, unique, and authoritative individuals (George 1994; Ghose 1998; Kapman 1997; Knapman 1986; Sharpe 1993; Ware 1992). While this self-defining process involves denigrating other Westerner travelers as tourists, it may also constitute them as menacing, as Western women search for uniquely 'free' identities.[8]

James Buzard (1993a) argues that the function of travel for Europeans in the colonial era was to realize a fuller sense of self. Janet Wolff (1993) supports my conclusion that this is still the case, especially for women who historically have had a marginal relationship to travel. The requirements of femininity involve sticking close to home, thus making traveling a masculine activity. By choosing to leave home, like Shirley Valentine, Western women in Gilgit disrupt discourses of femininity related to travel, and thereby enhance their self-definition as independent individuals. Their individualistic strategies of escape through travel both enact a discourse of the bourgeois self and initiate a form of gender power for women: entering a global world enables them to transgress gender norms at home; to shape a self-confident, somewhat elitist female identity; and to gain some sense of control over their lives. The feelings of independence and confidence Western women achieve from traveling helps them to realize other, 'freer' selves. These feelings also serve to dull my research participants' experiences of cultural dislocation by translating a general sense of unease and vulnerability in Gilgit (discussed in chapter two) into moments of self-assuredness. The discourse of Western women's freedom, then, is another means of coping with the experience of being 'out of place' in Gilgit (Cresswell 1996).

This self-definition through travel is enhanced as Western women in Gilglit define themselves in relation to imagined indigenous women, whose lives are seen—primarily through the discourse of the passive, oppressed 'Third World (invariably Muslim) Woman'—to be devoid of freedom and autonomy (Bulbeck 1998; Jolly 1993; Mohanty 1984; Spivak 1985). The process of knowing Gilgiti women's lives and realizing emancipated selves involves, as Mervat Hatem (1992) describes, Western women projecting onto racialized and 'culturally inferior' women what they most fear in themselves: their experiences of gender oppression in the West, such as their inability to find well-paid, full-time work and the overwhelming nurturing and domestic responsibilities that are pushing them away from home. As Western women in Gilgit enact discourses of the bourgeois self and the 'subjugated Muslim Woman,' they define themselves as independent subjects who are free to travel and work abroad doing philanthropic activities without constraint. By imagining Gilgiti women as an oppressed mass in need of their benevolence—or at least of their superior cultural example—Western women avoid confronting their own gender oppression at home. This characterization also allows them to capitalize on racial inequalities to realize personal freedom.

In what follows I outline some of the processes by which Western women's 'liberated' subjectivities are formed in relation to their experiences of travel and to 'oppressed' Gilgiti women. First, I discuss the dynamics of the relationship between Western and Gilgiti women, which are informed by intersecting discourses of gender, race, class, and Orientalism. The importance of clothing as both a sign of Gilgiti women's oppression and a site of subjectivity formation for Western women is the topic of the second section. I then discuss the relevance of international development activities to the constitution of their subjectivities. Work identities signal Western women's liberation and cultural superiority over Gilgiti women. Through these identities, my research participants also often draw on practices of cultural imperialism. Finally, I explore a major contradiction in the lives of 'free' Western women in Gilgit: the gender oppression they experience in Gilgit, effected as much by foreign men as by Gilgiti people, makes their efforts to construct identities as independent and unconstrained subjects even more necessary and fraught. Women often draw on discourses of race, class, and Orientalism to cope with gender discrimination and to reinforce a sense of freedom and superiority over indigenous women. These ambivalent experiences of gender freedom and constraint in Gilgit prompt some of my research participants to reevaluate their self-representations as 'free' women in the West.

Free Selves/Subjugated Others

When a crowing rooster interrupted our interview, Amanda shouted "I want to kill that rooster! Every morning he wakes me up at six . . . He struts around the garden, he's the boss, and all the girl hens follow behind him. Yeah, it's like the local women here [laughs]." By naturalizing Gilgiti women as a group of subjected followers, Amanda perpetuates two interrelated discourses of Orientalism that objectify and homogenize Muslim women as an oppressed mass. She also implies that it is only Muslim women who follow. According to the Orientalist beliefs prevalent among most of my research participants, Gilgiti Muslim women are radically different from Western women. Liberated Westerners usually contrast themselves favorably with oppressed Muslims. This liberated/oppressed dichotomy extends to related oppositions: progressive/traditional; uncovered/veiled; free to travel and work/confined by purdah and the veil; independent/dependent;

active/passive; unique/undifferentiated. This last opposition, which depicts Gilgiti women as a faceless mass, is rooted in a discourse of Orientalism that has historically legitimated Western women engaging in travel as a means to achieve personal uniqueness, as well as to knowing and representing homogeneous female Others (Lewis 1996).

Without prompting, almost all of my research participants painted fairly uniform portraits of Gilgiti women in our interviews, which seems to indicate a certain fascination with knowing them, even though that fascination, curiously, does not extend to learning about their everyday lives. As Amanda proceeded to discuss her maidservants' various health-related ailments, she launched into a more extended description of Gilgiti women's putatively onerous lives: "Here, the women are always suffering. They give too many children birth. And they let men handle them like weak people. That's their lot. They can't be seen or heard. That's a big problem here for me, the plight of Pakistani women." Gilgiti women's apparent lack of agency to fight their 'lot' is almost as bluntly represented by Andy, who was "brought up to know that I did have a say, and that I could say 'No!' to [men]. And I think women here *don't* have that. It was quite interesting when my girlfriends were out here. They were talking about women here who don't even know they have the right to say 'No.' Is it abuse or is it rape? Probably not, it's just *life* here." Christine, like Andy, problematically defines herself as superior in relation to inferior female Others by drawing on an ethnocentric and Orientalist logic. For instance, she complains that "There's this assumption here that, as a woman, you will just fall in with [men's] ideas. I mean, they're used to me now. They know that *I* won't, even if local women do." But at another moment, Christine, retaining a patronizing tone, seems ambivalent about how much freedom indigenous women have or can achieve: "Some [Sunni women teachers] can't leave for jobs outside the valley, because, of course, they're not allowed to travel and the men just can't be bothered to escort them everyday. So, blow me, instead of sitting by the fire and moaning, they've opened their own little school. It's just brilliant, just wonderful." In one instance Elena seems similarly unconvinced that all Gilgiti women face a uniform lifetime of oppression. Despite her sense that even the most happy local marriage cannot meet the Western standard of spousal equality, she is impressed with her work coordinator, who has ignored family pressure to take a second wife so he can sire a son: "They're *so* much in love, he and his wife. It's a *wonder* to see. They've got, I suppose, as equal a relationship as you can get here." But a few minutes later she laments that

> Women here, ahhh [sighs], they never get to go anywhere, and they're
> stuck in their homes all day, every day. That honestly breaks my heart,
> it does. I just think 'God, I don't know what I'd do.' You can't think that
> all these women are born like this, wanting to stay home and be humble
> and shy and quiet. There must be people like us [Western women]
> around who are extroverted. It's just pushed down. It's terrible . . . The
> way women live here is sad. Just to have *babies* and be servants to their
> husbands.

Although she trains Gilgiti women teachers, and thus realizes that
they experience possibilities as well as constraints, Elena ignores
that agency by perpetuating the view that Muslim women, *en
masse*, are confined to the male-dominated space of the home. Her
focus on the home as the primary site of Muslim women's oppression, like that of much Euro-American feminism, recalls Western
women's colonial-era fascination with representing Muslim women
subjugated at home in the harem (Chaudhuri 2000, 2002; Lewis
1996; Mills 1996; Shohat 1991; Weber 2001). It also sustains the
falsely homogenizing notion that Muslim women's concerns are
largely domestic, in contrast to Western women travelers whose
development work shows them to be well-educated, sophisticated,
and worldly (Hatem 1992).

Rose also locates indigenous women in the "traditional," unen-
lightened space of home, although she does so metaphorically.
Sometimes she feels she is

> with these women in a dark, traditional hut, where there's only one
> window that is always closed, and when I'm there I can open the win-
> dow. 'Look, there's something out there for you.' But then afterwards I
> have to close it again. And they must open it for themselves. But they are
> far too shy, because of the Islamic man-woman thing . . . But that's how
> I feel my role is in this women's group.

Abbie agrees that Gilgiti women are confined in their lives, but relates
that to her own experience in the West: "Local women have neighborhood groups, and they get input [into community life] that way. But
that's all they know. Maybe with a lot of them it's absolutely fine.
They're not mistreated, but their worlds are so small. Of course, then
you start thinking that some people's worlds in the United States are
awfully small too [laughs]. They've hardly been to the next town

even." By drawing a parallel between Western and Gilgiti women's lives, Abbie disrupts a strong free/subjugated binary.

While I did not ask specific questions about Gilgiti women during interviews, I did ask my research participants to describe themselves for me. And when they did, as some of their comments show, they almost always drew on discourses of Orientalism to characterize themselves as 'free' Western women who contrast favorably with indigenous women. For example, René is adamant that

> The *best* thing about Americans is their tremendous freedom. It stops my heart when I think about the freedom of America . . . Independence is a very Western thing and dependence is a very Eastern thing. For an American to be very independent can be good, as far as expanding their minds. And creativity is a very positive thing that comes from independence . . . Whereas dependence, with this extended family or cultural expectations, that can be very confining and keep a person from developing themselves and their skills, and having the joy of learning things . . . Dependence can be very oppressive and confining.

Marion thinks that "what local women envy about us [Western women] is our independence . . . our ability to live alone and to make our own decisions . . . and to make our own money. I point out to them that there can be *too much* independence, and that people can become very lonely and very isolated, but they, of course, like everybody else, can only see good in what they don't have yet."

Anguita, who is not a development worker and has not come to Pakistan for putatively philanthropic reasons, tells a story that counters this notion that Western women live free of oppression. She does not "know what a bad experience is in Pakistan. I certainly had worse experiences in Europe. I was raped in Yugoslavia once, because I was hitching. And I had a bad story once in Spain. If I bothered to look at it, I had more cases of abuse in the West than I've had in Pakistan. What can I say?" But, as Rose indicates above, other Western women in Gilgit see both their emancipated selves and their development work as sources of inspiration and redemption for Muslim Gilgiti women. Andy believes "You can't solve the world's problems. You can't fight all the battles. And I prefer, in a way, to work at empowering local women to do it themselves. But I sometimes make a point of walking through the bazaar wearing a t-shirt and pants, because I think if I don't, no local woman will ever be able to." Using an ethnocentric logic, Andy seems to assume that local women want to dress as Western women do, but are prohibited.

Likewise, Abbie hopes to empower indigenous women, through the example of her 'equitable' marriage:

> My idea of what I want to do here isn't so much trying to change local people, which some [Westerners] really are, and I don't think that's all bad . . . [My husband] and I think one thing we can show people that might be different is just our love for each other, a man and his wife, that is completely equal, working together, that love each other, so it's obvious. That must be an example for people here . . . I just think that probably none of them have that sort of relationship . . . It's good for them to see an equal marriage that is good and where the husband isn't defensive or anything.[9]

While many Western women hope to enable the emancipation of Gilgiti women in a similar way, Evelyn, in her usual pithy manner, contrasts herself to missionary NGO women who 'convert' by example: "One of the things for these Christian women is to know that they could be playing this game for the rest of their lives. I'm going to leave, and I will probably hurt some people's feelings, but luckily they won't have to *see me around* [laughs]. I hope they can forget about me." Evelyn's effort to distance herself from other women's conversion activities destabilizes dominant beliefs about Western women's superiority and shows that not all Western women agree about the best way to go about empowering Gilgiti women or making them anew.

The discourse of the 'free Western Woman' does more than perpetuate subjugating bourgeois, ethnocentric, and Orientalist ideologies; it also disrupts the discourse of racialized sexuality, which implies that safety is possible only when women are protected in their homes. As Indira Ghose (1998) demonstrates, this discourse marginalized Western women in the colonies when their male compatriots exercised it as a tool of patriarchal boundary maintenance to justify relegating them to home and regulating their sexuality. But when contemporary Western women wander outside their Gilgiti homes independently, they demonstrate that women can manage their lives without men's intervention, thus exposing some of the myths of racialized men's sexual threat. As I demonstrated in the previous chapter, many of my research participants perpetuate the discourse of racialized sexuality through their spatial, cultural, and representational practices, but their adventurous behaviors and active self-images simultaneously subvert it by undermining the premise of white women's vulnerability—a premise on which this discourse is grounded.

The binary liberated/oppressed leads most Western women to overlook characteristics, events, and situations they have in common with Gilgiti women.[10] Indigenous and foreign women's similar gender experiences and coping strategies may be too disturbing for many of my research participants to acknowledge, as any resemblance between 'us' and 'them' collapses cultural and racial divisions upon which their contemporary 'civilizing mission' is based. The freedom of Western women thus relies, in part, on demonstrating their cultural and racial superiority over Gilgiti women. The following quotations, which describe a lack of perceived connections between Western and Gilgiti women, illustrate how my research participants represent indigenous women in ways that problematically presume, perpetuate, and augment their own racialized sense of distinctiveness and superiority.

Andy, like some Western women in colonial India, does become acquainted with well-educated, wealthy Gilgiti women who are fluent English speakers (Chaudhuri 2002; Jolly 1993). But even her closest acquaintance cannot satisfy her desire for stimulating, intellectual conversations: "This is a horrible thing to say, but . . . down-country Pakistani women, their education is so much less effective than the men's here, that even the most educated women don't argue even on the same level as the men. They don't want to talk about the same things *I do*. Amina next door, she has her MBA and everything, but in terms of potential for challenging conversation, maybe about *clothes*." This remark is also notable as an example of how discourses of race have shifted in South Asia since the colonial era. Through feminist and multicultural movements in Britain, Andy knows this racist claim is unacceptable. She apologies for it, but ultimately cannot help express the idea. Foreign women living in colonial India usually would not have had such reservations.

Margaret stresses in an even more disdainful way the lack of intellectual connections she has with Gilgiti women. Her "friendships here with local women only go *so far*. Intellectually, people aren't stupid here, but we [Westerners] are light years ahead [of them]. We're on a completely different planet . . . You don't realize how *skilled* you are [until you] live and work here." Evelyn more subtly describes why she is reluctant to cultivate relationships with Gilgiti women outside her workplace: "I see local women at work, but I don't see local women socially very much at all, and it's not because I don't like them. It's because I don't want to use my discretionary time that way to be truthful . . . Language is not always a barrier, it's disposable time. And also feeling that a [visit] would be a very formal, lovely thing, but it wouldn't give me what it is I need in terms of *connections*." Due to their tendencies toward dependence, Susan thinks she has "nothing in common with [the neighbors]. There's a woman who lives across the back here . . . At the beginning I was quite friendly with her, but then she started to . . . ask for things . . . But I don't like that idea. It perpetuates dependence . . . .And that's not what friendship is about. So, I found that I don't have anything in common with most neighbors."

In contrast to these women, Julia has made an admirable effort to develop relationships with Gilgiti women at work by visiting them in their offices and joining them at tea and lunch breaks. She told me that, once her Urdu improved, she tried to find commonalities with these women so she could develop some sense of belonging in Gilgit. My own desire to live with indigenous families in Gilgit (after I confronted my Orientalist preconceptions about Muslim women) was directed, in part, by a similar aspiration: to learn about the everyday lives of Gilgiti women, find a niche for myself, and understand local rules for gendered behavior. Louise also cultivated close relationships with the senior women in her host household. But, ironically, even her experiences of gender oppression, which overlap with those of indigenous women, serve only to reinforce the free/oppressed dichotomy:

> Being involved in family domestics has caused terrible problems for me, because I'm being dragged into arguments between husband and wife . . . Just the other day . . . I didn't really want to be washing up, but I was . . . It's [men's] habit to [yell for water]. And I was washing up, and then came "Louise, Louise, water!" And I took a sit back for a second and thought "Here I am, quite confused really, because one minute I'm this Western woman who goes off to work independently, but now I'm here, washing up and taking on the role of a local woman." And all of a sudden I had a panic that I was starting to be seen [like that]. So I ignored the request, and it came again, and then he said to his wife "Water!" . . . I got cross and said "Can't one of you come and get it yourself? It's just here." And I had a little go about it, and then the wife, from nowhere, just exploded [laughs]. I'd never seen it before . . . He said "You're bringing your Western views into our house, and trying to change it." I took a big step back and thought "I *can't* do this. I can't demand that this happens, but, at the same time, I can't deal with it *inside* . . . Where can I draw the line? Can I honestly say this is right for me to step in?" . . . Faridah has to carry on . . . Will my talks make life difficult for her? I don't think it did. I just gave her the example, perhaps gave her the confidence that, as a woman, I was speaking out. I don't know what quite triggered it, but I think my presence has helped it.[11]

Louise's story reveals that a Western woman can have interactions and develop friendships with indigenous women, can experience the same gendered expectations, and still not acknowledge having anything in common with them, even domestic duties such as washing up. This oppressive act of denial suggests that Western women can be threatened by the prospect of being anything like indigenous women. Congruent with a subjugating 'West is best' attitude, an unbearable likeness of being would undermine their sense of racial and cultural superiority, which is perpetuated as they draw on discourses of race to claim intellectual capacities, independence, and imperial authority.

Once they decide Gilgiti women are unsuitable companions, my research participants must search among themselves for friendship and intellectual connections. But this search is not always easy, even though Westerners often have similar cultural traits, including language and tastes, and a craving for cultural belonging. The expatriates in Gilgit form what Liisa Malkki (1997) calls an 'accidental community,' a group that participates in a *transitory* experience and a *memory* of cultural belonging, as opposed to a sustained, manifest, and voluntary engagement. In this way, they are like refugees in a makeshift camp. The transitory and accidental aspects of their community have significant implications for how friendly foreigners can be amongst themselves. Although they speak the same language and enjoy similar foods, drinks, and music, being thrown together haphazardly through their development work means that they often have little else in common: they haven't read the same books, their sexuality and spiritual beliefs differ, they range widely in age, they are not all parents, and they espouse various and competing philosophies of life. According to Susan, "I have made a life for myself without getting too attached to the other volunteers. We're friendly, we socialize, we go out together, but   people are quite distant, we don't have much in common . . . I have a [marital] relationship that's very strong, so I'm lucky. I don't need to rely on anyone else . . . Then you can accept that sort of fair-weather friendship . . . But it's very superficial for me." The fact that these relationships are temporary also affects how deeply Western women are willing to be committed to one another, how much they will invest in intimate friendships, because emotional attachments make leaving more painful. Janet has learned to be

> more independent about relationships. It's nice to have people around, but . . . When *you* went, quite honestly, we had talked *such* a lot, *such a lot*. We'd also shared social occasions together . . . So it is a little *death*, it really is, a little death when somebody leaves. And in this kind of harsh environment, where you feel lucky to find some kind of rapport with people that you're with, it is quite definitely a little kind of death.

Many Western women in Gilgit attempt to prevent emotional pain by not getting too close to other foreigners. For instance, Abbie is "hesitant to get too close to anybody. I am not going to carry other people's burdens up here too much. I'll pray for them and I'll do what I can, but in the early years, we just felt like we had to be there for everybody . . . Not everyone has what it takes to live up here." As a result of remaining emotionally detached, some of my research participants feel alone and unsupported. Andy states,

> I started off in the VSO crowd, and I moved away from it because I find no support there, for lots of reasons. I think the sort of person who signs up for VSO is exceptionally independent, hard-headed, contained, and they don't need support in the same way, so they're not as capable of giving it . . . I've needed support from friends [when the job isn't going well], but you don't have that depth of friendship with people that can actually support you when you need help. Because people are living on quite a superficial basis, they're not emotionally involved with you . . . You're sort of a bit nothingy to them.

Although Western women do not often experience relationships with their compatriots as gratifying, they do everything they can to prevent tensions building between workmates, housemates, and socializing companions. These associations must be kept intact not only to create sustaining human contact, but also to maintain the comfort and illusion of Western sameness/Gilgiti otherness.

The transitoriness of this expatriate community also influences my research participants' commitment to learning a local language. Only a few Western women in Gilgit are interested in speaking Urdu, Shina, or Burushaski, yet most of them hope to live or travel in the area for many years.[12] Abbie took an 18-month Urdu training course in Muree before starting mission work at the eye hospital. Julia has learned to speak Urdu well, and finds it puzzling that other foreigners can survive without it. Jane and Dolly are taking private language lessons, and Fiona regrets that the Urdu she learned in a Multan mission is of little use when working with women from the Gilgit region who speak only Burushaski. The rest of my participants admit they have very poor language skills, in consequence of being in Gilgit for only two years. As I do, most of them find learning new languages difficult. And they feel that the time and effort required to speak a new language is wasted on such a short visit. Moreover, VSOs are overworked on their two-year contracts, leaving them too tired to pick up another language on their one day off a week. And as most of them socialize exclusively with other Westerns and work with indigenous people who are trying to improve their English, they think learning Urdu is not necessary. While these are plausible, practical reasons explaining why my research participants have not acquired a local language, there are other, more troubling, ideologies at work supporting this resistance. Marion told me that

> People here don't want to speak anything but English to us, even in the bazaar . . . So I haven't found any need for Urdu. And not only do teachers not want us to speak anything but English, I feel I would be letting them down if I spoke anything else, because their English is so much in need of improvement that every opportunity they get should be maximized . . . If I really felt it was important, I could learn Urdu. I quickly learned, for example, all the names of the fruits and vegetables, because when I go to the market I need them. And I can do general greetings and politeness, but that's all I need.

Here Marion exercises ethnocentric discourses to cultivate a sense of cultural superiority in two main ways. First, she implies that English is a more consequential language than Urdu, even in Gilgit. Second, she characterizes Gilgiti teachers, who instruct children in English, as poor English speakers and herself as a quick study in languages, when she's inspired. But she is not particularly interested in learning Urdu, because it is required only to converse with indigenous people. When Marion anticipates having only superficial interactions with Gilgitis in the bazaar and perhaps in school, learning Urdu is not a priority. This lack of interest in learning a local language, which would facilitate more constructive Western-Gilgiti interactions, is also evident in the syllabus of the expatriate public school. The American home-schooling curriculum has been revamped to include religious, but not Urdu instruction. When I asked the teachers why the children were not learning Urdu, they explained that language lessons would take time away from more important subjects, like English. Moreover, the children, who play almost exclusively with each other and leave Gilgit for English-based high schools down-country or abroad, have little need for Urdu. When Westerners restrict their interactions with Gilgiti people by not learning a local language, the alienating social and cultural distance between foreigners and locals persists. Gilgiti women, who speak English far less often than Gilgiti men, are especially detached from Western women in this regard.

Although my research participants are ambivalently Orientalist, they most often see Gilgiti Muslim women as radically different from

Western women. Thus, by viewing indigenous women through an imperial lens as 'oppressed,' they can perceive themselves as 'free.' In this way Gilgiti women are trapped in Western women's lingering, although ever shifting and fractured, colonial imagination. And in the process, discourses of Orientalism—particularly the notion that Western women are culturally and racially superior to downtrodden, submissive Muslim women—are perpetuated. Specific Orientalist practices also have troubling racist implications. For example, when Western women purposefully avoid interacting with Gilgiti women, presume they have nothing to learn from these 'intellectually underdeveloped' and 'dependent' women, and impose their cultural norms and practices on 'ineffectual' locals, they promote oppressive discourses of race and of the 'Orient.'

My research participants' ambivalence toward Gilgiti women is often expressed when discourses of bourgeois femininity/domesticity and Orientalism intersect. As Christian missionary women perceive themselves as 'angels of the bourgeois hearth,' they sympathize with Gilgiti women's seclusion in the home and the apparent domestic focus of their lives. Whether they crave privacy from indigenous people or feel that their Christian duty includes being respectable wives and mothers and keeping well-managed households, these women have daily routines that keep them close to home or to the expatriate public school where they teach their children. René wants "the kids to have a good education and have lots of good experiences, so I put my energy into that . . . My day is very much wife and mother directed. So, half of the time I'm teaching school . . . And then it's managing the household, making sure that we have food, and the dog is fed, and that I cook supper, all the domestic needs." And Jane is "very concerned that women and their behavior affect the way the family is perceived. And I certainly don't want my behavior to do anything judgmental to [my husband's] position and being accepted by communities up here . . . So I don't wander outside our property . . . I keep my hair longer . . . I always cover my head, always, always, always." As it did in colonial India, this adherence to the dictates of femininity separates Christian women from other Western women in Gilgit, who have not been pulled to Gilgit by husbands and a sense of familial duty (Melman 1992; Nair 1990). The discourse of bourgeois domesticity conflicts with discourses of the self that are exercised through emancipatory travel experiences and representations of Gilgiti women as the radically Other. It also undermines the notion that Gilgiti women are oppressed because they are confined to the home, as these foreign women's lives are similarly domestic. Some sense of reciprocity and solidarity with indigenous women can disrupt discourses of race and Orientalism. When Abbie argues that confined Gilgiti women make important community contributions and that they are not mistreated simply because they spend most of their time at home, she shows that there are significant discursive tensions between different groups of Western women in Gilgit. But ambivalence is manifested within, as well as among, subjects. Christian women's sympathy for Gilgiti women's lives conflicts with their feelings of racial, cultural, and even moral superiority over Muslim women. Abbie may have some empathy for Gilgiti women's domestic lives, but she believes her love marriage, as a sign of Western cultural superiority, is a standard to which indigenous men and women should aspire.

These ambivalent perceptions of Gilgiti women show that there are different metropolitan 'gazes' employed to create Other women as the polar opposite of Western selves. However, each act of vision and knowledge production, no matter how conflicted, is part of a technology of power that disciplines Gilgiti women. As Franz Fanon (1967) and Denean Sharpley-Whiting (2001) argue in analyzing the 'white male gaze' turned upon black women, Western women's scrutiny of Gilgiti women consists of a desire to unveil the unknown, to make the latter group less opaque, to fix them in their place so they can be translated into a Western system of representation, and thus managed. A desire for knowledge of difference can be a desire for mastery, but it can also signify sublimated fear, fear of racial and cultural difference, as well as Western women's fear of confronting their own gender oppression (Nair 1990, 25). Many of my research participants maintain a sense of cultural and racial superiority by projecting their concerns for personal freedom onto Gilgiti women, a process that makes Western women appear 'free.'

What to Wear, What to Wear?

In this country, Western women's identity is often focused on the *chador*. Wear it or not wear it. Resent it. Wear it with resentment. Hate it and express all that elsewhere. And a lot of that is the focus on that *stupid* piece of cloth. And it must be a lot of other things, actually, being verbalized in that one thing.

—Jane

When Western women in Gilgit understand themselves—in contrast to tourists—as aspects of the local landscape, they become associated with Gilgit's state of 'cultural atrophy.' Gilgiti women's clothing is one cultural site where this decay is signified and managed by my research participants. *Shalwar kameez* is understood as the appropriate clothing option for all women (but not for men) in Pakistan, because it enacts and respects Islamic codes of modesty. While culturally detached tourists usually remain in their Western traveling clothes, my research participants' line of "cultural discretion" (Clifford 1997, 75) includes wearing this garment to signal their understanding of and respect for local culture and people, and to win personal respect in return. However, *shalwar kameez*, with its accompanying *dupatta* or head-covering *chador*, is also seen by many Western women, through an Orientalist lens, as a sign of Gilgiti women's oppression, an imposed (and ugly) form of chastity that makes women ashamed of their bodies. Thus, the decisions 'free' Western women make about what to wear in Gilgit are salient and conflicted practices of subjectivity formation, which are informed, once again, by Gilgiti women's behaviors.

Within the scant anthropology of clothing literature, authors seldom explore the overlap between clothing choices and processes of identification and differentiation.[13] In contrast to ritual or 'fashionable' dress, the topic of everyday clothing—except in regard to its design, production, and distribution—is often sidelined in ethnographic studies. Feminists in particular may avoid the topic as a stereotypical example of fetishization for women. But I want to demonstrate here that clothing—an embodied set of cultural codes—is a vital part of performing subjectivity and enacting power. It is "one of the most visible markers of social status and gender identity and therefore useful in maintaining or subverting symbolic boundaries" (Crane 2000, 1). By outlining how Western women in Gilgit negotiate the problem of what to wear, I illustrate how they use the shifting and disputed meanings of both Western outfits and *shalwar kameez* to express and manage their identities. I also show how these expressions of identity display important cultural values, in particular the Western bourgeois values of freedom and individual choice. Women decide what to wear by negotiating Western and local rules about clothing, but also in relation to their actively constructed self-images, making these decisions a complex amalgam of individual 'choice' and social 'imposition.' As Emma Tarlo (1996, 8) argues "Clothes are not merely defining, but they are also self-consciously *used* to define, to present, to deceive, to enjoy, to communicate, to reveal and conceal." Clothing, as an extension of their selves, pronounces who they think they are and differentiates them from who they do not want to be (Wilson 1987).

To explore the role of clothing in constructing identity, we need to answer the question of what particular types of clothing mean to the people who wear them. For many of my research participants, Western clothes—particularly blue jeans and t-shirts—when they are worn in Gilgit signify their modern, independent selves, as well as their freedom of choice. When I asked Jean how she feels when she exchanges *shalwar kameez* for Western clothes at an expatriate party, she replied

> Great! Absolutely great. It's not awkward at all. I find it a release to get out of [*shalwar kameez*], and then putting the [Western] ones back on is just like the status quo, the norm . . . They feel much more relaxing. I feel like I know who I am inside Western clothes. When I put on a pair of jeans and a t-shirt I feel like I am this confident, independent woman, walking around in clothes I choose to wear . . . You need breaks [from *shalwar kameez*] realistically, to carry on.

Congruent with my analysis, Anguita interprets Western women's preference for jeans over Pakistani clothing styles as an 'obsessive' act of identity formation: "Western women here have a fetish with jeans. Sacrificing the jeans [for a *shalwar kameez*] is like sacrificing a piece of them, something deep from inside themselves." Although I have never packed jeans for a trip to Pakistan considering how awkward I would feel wearing them anywhere in Gilgit, they seem to enable many of my research participants to regain a sense of comfort and relaxation in an unsettling environment by expressing a familiar Western identity. For Margaret, wearing pants and t-shirt after work means that "I still basically adopt my Australian *soul* in the privacy of our own home . . . At the time I had pants on and a t-shirt because I was just sitting in my house and I wanted to *relax*." Louise never thinks "about it during the day, but when I come home and I'm the *real* me, that's when I want to put on some Western clothes . . . That's what I know, that's what I've lived in, plus I haven't *chosen* [*shalwar kameez*] . . . Pakistan chose it for me . . . I don't like that I *have* to wear it. I haven't chosen this. I didn't say 'Oh, I like that fashion.'" By choosing to wear Western clothes as part of their presentation of self, my research participants hope to signal their identity as liberated modern women who experience lives full of choice, even despite the risk that Gilgiti people may 'misinterpret' this representation as a sign of moral laxity. Expressing a modern, independent identity is especially important in places of cultural belonging, such as their homes,[14] and at times of cultural allegiance; mixed dancing to British pop music is hardly imaginable without jeans and t-shirts. Consequently, and as Jean, Margaret, and Louise suggest, shifting from Western dress to *shalwar kameez* often feels like a partial renunciation of self.

In contrast to jeans and t-shirts, many Western women in Gilgit, like British women living in colonial India, see indigenous women's clothing as a sign of oppression (Chaudhuri 2000; Procida 2002; Ware 1996). My research participants, therefore, imagine Western clothes to express cultural superiority, as well as cultural belonging and fidelity. A shift from jeans to *shalwar* signals an embodied passage from a modern, civilized Western culture to a 'backward' Islamic one. According to Marion,

> The *dupatta* is just a useless piece of material that does nothing except get in the way. So I just hate it, chiefly for practical reasons. Symbolically I hate it too . . . It's a case of keep yourself hidden, you're a woman. If men give way to their uncontrollable lust, it's certainly not their fault. It is the women's fault, and therefore keeping yourself covered is required. And I don't actually agree with that particular anti-quated view [laughs] . . . The *dupatta* to me is the symbol of everything that is repressive for Muslim women.

Marion's description of indigenous women's head-coverings recalls a passage written by Florence Nightingale, who in 1849 disguised herself as an Egyptian woman: "I felt like a hypocrite in Dante's hell, with the leaden cap on—it was like hell to me. I began to be uncertain whether I was a Christian woman, and have never been so thankful as being so as since that moment. That quarter of an hour seemed to reveal to one what it is to be a woman in these countries . . . God save them, for it is a hopeless life" (quoted in Melman 1992, 103). Both quotations manifest what Stoler (1995, 104) calls "colonial anxiety," Westerners' fears that elements of the colonized territory, such as clothing, have the cultural power to change them, to threaten their identity through "native contamination."

Despite Gilgit's postcolonial situation, troubling cultural stratifications between Gilgitis and Westerners are pronounced enough for *shalwar kameez*, particularly the *dupatta*, to signify Muslim women's ostensible subordination. As Marion shows, some Western women in Gilgit see Pakistani clothing styles as 'primitive' hold-overs from an oppressive past forced on Gilgiti women, rather than as a dynamic cultural form. For example, Lyn, ignoring indigenous women's agency, thinks that "by wearing *shalwar kameez* I'm actually, in my own head I'm actually *supporting* the oppression of women in this community, because they're *still* being told they *must* wear this. They have no choice." Jane enjoys wearing *shalwar kameez*, by choice, as a sign of her devotion to Pakistan, but imagines it means something very different for Gilgiti women: "Generally you think of situations here where oppression is the lowest common denominator, in terms of the way local women must dress and they don't get to go out here. They don't have a lot of opportunities or choice." There is, of course, a fairly wide spectrum of metropolitan responses to *shalwar kameez*, which I discuss below, including some appreciation of the garment that denies the superiority of Western clothing styles. However, the dominant perspective is based on a disturbing stereotype that posits Western fashion as superior, Gilgiti styles as proof of an unchanging 'primitive' Islamic culture, and Muslim women as oppressed victims of fashion who have no choice in what they wear. *Shalwar kameez*, then, facilitate the marking of difference between Self/Other, modern/primitive, and autonomy/dependence.

Although most of my research participants see *shalwar kameez* this way, some do not. The women in this group are British or American, and vary in age (from 36–50), partnership status, religiosity, and length of stay in Pakistan (1–5 years). Characteristic of this group, Joan

> love[s] *shalwar kameez*. I think I had 30 last time I counted . . . I just love finding the fabrics and the colors and having them made. I know there's *no* variety [in design], but I find that they're *great* fun . . . I love clothes, so I've enjoyed them . . . I think it's possibly the *change* for me, because I'm very much a suit with being the head of a school for so long . . . And I find them comfortable to wear. You can sit any way, can't you? And you don't have to wear tights. There's sort of a *freedom* about them, isn't there? . . . *Practically*, the *dupatta* is a problem, but I don't have any personal problem with it. I don't feel it's unnecessary. I do think it *finishes* the outfit often, doesn't it?

Joan's love of fabric, fashion, and flair stimulates sympathy for *shalwar kameez* free of any sense of cultural superiority. However, by denying the ever-shifting details of its design, while simultaneously not noting the monotonous motif of Western business suits and jeans with t-shirts, her characterization of the lack of variety in *shalwar kameez* subtly recuperates just this implication. She delicately exercises an Orientalist discourse that positions a static 'Oriental' culture in contradistinction to a flexible and dynamic Western one (Said 1978).

I have a similar fascination with the seemingly endless quantities of beautifully colored cottons and silks sold in the bazaar, and I agree with her that *shalwar kameez* are comfortable to wear and incomplete

without a *dupatta*. Although I devote much less attention to my clothing at home, the challenge of styling and wearing *shalwar kameez*
fashionably preoccupies me in Gilgit. Julia and Jane express a similar
sympathy for the outfit, but primarily as a means to achieve cultural
belonging and respectability. Jane "like[s] *shalwar kameez* . . . I've
had women say to me 'Oh, you're dressed properly' . . . If I was wearing shorts and a skimpy t-shirt, they wouldn't be able to talk to me,
but they can associate with me because I'm dressed sensitively . . . I've
never wanted to send any sort of ripple out, for my husband's
sake . . . So I decided to err on the conservative side, because people
always remember [a woman's] behavior." Jane's *shalwar kameez* may
mark her body with her desire for cultural belonging and respectability, but her nose pin does more. It is also an embodied manifestation
of her experiences of Pakistan, an aesthetic symbol that marks her as
a sympathetic insider who can be understood in local terms: "I got my
nose pin here in 1985 . . . I thought it looked beautiful . . . But I didn't
realize how one little thing could affect [Westerners' good] opinion of
you . . . This is part of who I am. I live in Pakistan. I love the country.
And I wasn't going to not wear it because people couldn't see beyond
it." René has also signaled her experience of and sympathy for
Pakistan by incorporating local aesthetics onto her self-presentation:
"I had my nose pierced in Islamabad . . . I kept eyeing Jane's and
thought it looked nice. She had only positive things to say about
hers . . . I think they're beautiful . . . It's very interesting that *a lot* of
people [in the United States] ask me about it . . . I always tell them that
I live in Pakistan, and that in Pakistan it is considered very beautiful,
and that's why I did it."

While these women's sense of confidence, individuality, and autonomy seems to have been enhanced somewhat through Pakistani aesthetic experiences, a second group of women—who are equally
diverse, and range in age from 45 to 55—claim to be able to choose to
wear *shalwar kameez* without it affecting their sense of self. Janneke,
for example, does not "feel any different [wearing *shalwar kameez*]
actually. It's just a long blouse and a pair of pants. I don't feel any less
or any more European, or any more or any less Pakistani. I just feel a
bit less conspicuous downtown, but I still stand out." Dolly wonders
"Why should I worry about whether my image suits either Western or
Pakistani clothes? I know who I am . . . I don't have any problems
with [*shalwar kameez* and *dupattas*], because I'm still me. And at my
age, I'm just lucky that it's all, you know, still there [laughs]."
Rosemary argues that "If you're suitably covered, local people respect
you, you respect them, their country, religion, faith, culture, so why

not do it? Not only that, the clothes are cheap and comfortable. I don't wear Western clothes here . . . My body isn't that good to look at any longer anyway, so it doesn't really matter." Older women, such as Dolly, Rosemary, Janet, and Auguita, are also more comfortable wearing a modest *shalwar kameez* than younger women, because it hides bodies that are no longer taut and approximates the baggy pants and elongated shirts they usually wear in Europe, Britain, and Canada.

A third group of my research participants dislikes *shalwar kameez* as a sign of cultural primitivism and constraint, as well as of women's dependence, lack of choice, and shame. This varied majority range in age (from 23 to 59), nationality (American, British, European, Canadian, and Australian), religiosity, partnership status, and length of stay in Pakistan (1–7 years). Despite their collective dislike of *shalwar kameez*, almost all of them wear the garment, at least when they are out in public. Why would modern independent women choose to wear clothing that signifies what they see as local women's oppression? How do they experience, tolerate, and manage this partial renunciation of self? This ambivalence is required to garner the respect they need to accomplish the good work of development, even when they experience it as a disconcerting loss of self. Abbie explains that "Western clothes are more like me. I wear *shalwar kameez* to be polite to their culture for my *job*, and that's fine. But I don't really like it. I don't wear it because it's comfortable, because as soon as I put on Western clothes I feel more me, more who I am . . . I never came with the intention of becoming native [*sic*] or fitting into local life, so I don't feel like I've failed myself in that." To Susan, "[*Shalwar kameez*] feels like a prison uniform . . . But I would say that 95 percent of the time I'm in *shalwar kameez* as soon as I walk out the door. I think it's proper and professional to wear it. [Gilgiti] people like it, they do respect you for it. So that's enough for me. I'll wear it all the time, because it's important for work to get respect." The different representations of *shalwar kameez* circulated by these three groups of women demonstrate, in concert with Indira Ghose's analysis (1998), that the female colonial gaze is ambivalent and fractured. However, each distanced, 'objective' perception monitors and controls the Other through epistemic mastery, by (re)producing imperial knowledge about the Other while effacing its own implication in the specificity of power. We see this process at work in Susan's quotation, in her detached evaluation of *shalwar kameez* as undifferentiated 'uniforms.' While both Pakistani and Western clothes are uniforms, in that they are by and large practices of conformity, by ignoring this orthodoxy Western women can create a binary view of *shalwar kameez* and

Western clothes, and thus imagine, drawing on Orientalist knowledge, that Gilgiti women are oppressed because they lack the autonomy Western women have to choose from numerous styles of clothing.

While Abbie and Susan stress the professional benefits of winning respect by wearing Pakistani clothing, many Western women in Gilgit are also concerned with the personal benefits. They think, invoking discourses of racialized sexuality, that wearing *shalwar kameez* signals their modesty and respectability to Gilgitis, which will hopefully decrease incidents of sexual harassment by Gilgiti men. René, for example, feels "a very big need to have a good reputation. So to follow their rules a bit enables me to have that . . . So I just try to be conservative, reasonable, and careful so I have respect. I don't want to deny my American identity, but, at the same time, I want respect so I don't get hassled. So I do things that for me are fine, like wearing *shalwar kameez* and a *chador*." Evelyn has a much lower tolerance for this ambivalence, but she remains concerned about presenting a modest image for the sake of good work:

> I say that I despise *shalwar kameez* . . . They send me this message that my body should be covered, and I think it probably *should* be [laughs], but for different reasons than they think. I *hate* uniforms, I *hate* being told that somebody else knows what's good for me . . . I wore them *every day* for a year and a half, and then one day I came downstairs in a pair of jeans . . . But I think I had played the game long enough that people realized that I was a *good, modest* woman . . . I guess I don't look as bad in them as I feel, because I feel like *shite* in them. They're *sacks!* I don't want to wear curvy clothing. It's just this formula. I just want to wear what I want to wear . . . But I don't know if [wearing *shalwar kameez*] gets to my sense of myself or not. I don't know if it penetrates, or if it's just something superficial . . . I do it, because in situations where I'm not known I want people to think of me as a modest person . . . I always thought that the reason I was doing it was so I could make my entry professionally. I didn't want to compromise what I really *believed* in. I care about this work, so I'll wear it . . . .But, I've said it before and I mean it *absolutely*. If I had to cover my head, I'd be on the next flight out, *absolutely*, no questions asked.

Bulbeck (1998, 30) speaks eloquently to Evelyn's comments about being 'overcovered' by an undifferentiated 'sack': "When Western women were 'civilized' Victorians because they covered up, exposure was deemed barbaric. Today the covered woman has replaced the exposed woman as the signifier of the Other, indicating Western woman's superiority."

While this group of women may appear content—perhaps resigned—to wear *shalwar kameez* in Gilgit for the personal and professional benefits it brings, they also experience an ambivalent imposition of a supposedly homogenizing and oppressive Islamic culture over their unique and liberated metropolitan bodies. They experience it as altered body images, infused with feelings of bodily shame that are at odds with notions of Western sexual liberation. Louise has "become a lot more aware of my bust, *really* aware of it, and *very* conscious when I'm in the bazaar to cover up. I'm now quite big chested. I used to be proud of that at home, but certainly not here." Andy is "body conscious in England, but never to this extent. Not so that when I wear a seat belt that comes down here I put a *dupatta* over the top so you can't see my breasts. I mean, what am I doing? Or when I wear a t-shirt I feel quite self-conscious . . . And then Dan says 'Well, that's a bit of a revealing top.' I mean, it's a t-shirt!" When Susan was on a VSO funded furlough in Sri Lanka,

> it took three days for me to get out of a long-sleeved shirt [laughs] . . . On the first day on the beach I got into shorts and a t-shirt— I had my bikini on underneath—and it took me a while to take [the shirt] off . . . It took me a few days to adjust to the fact that it was alright, that no one was going to stare at me or hassle me, no one was going to be offended, that I wasn't being *rude* by taking my shirt off . . . The idea of going [to Britain] in the summer and seeing all this exposed flesh is a bit too much for me now . . . I love it there, but the idea of what's normal in London would be quite shocking. I think it would be a huge culture shock. I probably wouldn't go out for a week.

Over time, Western women wearing *shalwar kameez* can amount to embodying the identity of a Muslim woman, with all the constraints and benefits that such a thing implies. To preserve some sense of their Western identity, these women shift to Western clothes in their homes and at parties, once again, as a way of defining themselves against Gilgiti women and asserting cultural distinctiveness in places of cultural belonging. For my research participants, Western clothes thus feel, in a psychic and physical sense, more comfortable and less restrictive than *shalwar kameez*, which allows for different, more 'normal' behaviors and physical movements.

Some Western women in Gilgit also negotiate this partial renunciation of self by altering their *shalwar kameez*. Small changes in style allow them to personalize the 'uniform,' mark the garment and their bodies with Western values of freedom, choice, individuality, and notions of beauty, while complying with indigenous norms of modesty and respectability. They devise a compromise between how they think other people want them to dress and how they wish to dress. René manages by wearing "Western dresses with *shalwar* underneath, because I feel like it's more attractive [than proper *shalwar kameez*], and I feel better about myself when I feel a bit more attractive." Susan doesn't "have *shalwar* made anymore. I have normal trousers, so I can wear them separately . . . That works for me, because I am compromising. I'm wearing *shalwar kameez*, I'm covered up from top to toe, but I can still wear trousers. I can take off the *kameez* and feel all normal." When Margaret first arrived in Gilgit, she made

> an extra special effort [to wear *shalwar kameez* appropriately], because you want to be accepted, and you have to be accepted to be trusted to work effectively. So you really go out of your way. But I suppose you get to a point, and I think about right now with you, I'm putting my *dupatta* on the table here instead of wearing it as you are. And I have this one *shalwar kameez* that is slightly adapted . . . The *shalwar* are a bit more like pants, which are *incredibly* comfortable . . . You know, you have to have *one* thing that sort of breaks the rule a little bit.

Although I myself am careful to cover my chest with a *dupatta* in public, my clothing rule-breaking consists of replacing long *kameez* sleeves with short ones.

As Margaret implies, compromises are most frequently made around the dreaded *dupatta*. Altered *shalwar* can be comfortable, but most of my research participants practically, philosophically, or psychically struggle with this chest-covering piece of cloth. Some women deal with it by wearing it over one shoulder instead of hiding their breasts. Sometimes they remove *dupattas* entirely in the office or classroom, for ease of movement. Shorter lengths of cloth are also an option. But the most pronounced deviation from the 'official' local style has to do with head-covering. Very rarely, and *never* in their homes, do my research participants cover their heads with a *dupatta* or *chador*, even when they like wearing *shalwar kameez*. As Evelyn declares, Western women devise ways to rationalize and manage wearing *shalwar kameez*, but the bodily shame and lack of autonomy supposedly implied by a Muslim woman's head-covering are attitudes even most cultural sympathizers are unwilling to accept as part of their identity in Gilgit. Some of my research participants would sooner leave the country and forfeit their work than cover their heads. Julia, in a commendable show of solidarity with and respect for indigenous women, does it in conservative villages where making herself more comprehensible to local people enables her to meet and interact with indigenous women. I have only covered my head twice, once while I was visiting an American friend doing research with devout Shia women in neighboring Baltistan, and once while traveling off the KKH through tribal Kohistan. Only Jane and René, two Christian women who identify more closely with Gilgiti women than my other research participants due to their religious and familial sensibilities and semi-cloistered lives, always cover their heads when they leave the house. According to these women, head-coverings, like *shalwar kameez*, symbolize their respectability.[15] A good reputation is vital for these women who plan to live in the area for some time and who feel that 'unsanctioned' behaviors can reflect badly on their husbands, and thus negatively impact their ability to do community development work. But for the rest of my research participants, this degree of respectability, which is ostensibly forced on Gilgiti women, is too costly to their sense of themselves as modern, liberated, metropolitan women.

Finally, what are the consequences of Western women's decisions about what they wear in Gilgit? First, I think that my research participants derive some sense of agency by asking themselves 'Is this *shalwar kameez* really me?' They hope that *shalwar kameez* will define them as morally acceptable and culturally appropriate to indigenous people for the sake of work, family, and personal safety. But they also self-consciously use both Pakistani and Western clothing styles to deal with their sense of alienation, and to *define themselves* as respectable, yet free, Western women.

Second, there are additional, less personal consequences to the notion that *shalwar kameez* defines identities regardless of how it is manipulated "to enjoy, to communicate, to reveal and conceal" (Tarlo 1996, 8). Western women in Gilgit have an ambivalent relationship to *shalwar kameez*. Those who enjoy wearing the outfit, or at least do not find it threatening to their identities, tend to be a varied group of older women who love fashion, seek personal respectability, or are familiar and comfortable with a loose clothing style that effectively hides aging bodies. In these two groups, then, age seems to moderate discourses of race and Orientalism that posit Pakistani clothing as 'backward' and oppressive. However, the diverse majority of Western women represent *shalwar kameez* as a sign of Muslim women's oppression, which means that they often use clothing styles to help define and naturalize hierarchical differences between themselves and Gilgiti women. By perpetuating dominant discourses of race and culture in these representations, as well as the associated subordinating subjectivity binaries of Self/Other, modern/primitive, and liberated/

oppressed, the *shalwar kameez* becomes the contiguous basis and proof of white women's moral, racial, and cultural distinctiveness. Like their spatial practices, Western women's clothing choices develop into arenas of inclusion and exclusion through processes of identification and differentiation.

Working Women

Development work is another important site of subjectivity formation for Western women in Gilgit. My research participants understand these activities as practices of philanthropy (according to discourses of femininity and bourgeois morality) and as a sign both of their freedom to travel as independent subjects and of their intellectual and cultural superiority over Gilgiti women. The discourses of race and class that sustain the representation of Western women as liberated, unique, and superior selves in and through the world of work are once again employed by my research participants to differentiate the latter group from Gilgiti women. But once these discourses become enmeshed in 'benevolent' practices of development that are meant to 'improve' indigenous people and sociocultural systems, some of my research participants also (re)enact imperial activities, while others challenge them.

Before I analyze processes of subjectivity formation through development work in Gilgit, I want to return to one of the 'push' factors that drove away many of my research participants who were engaged in paid employment in the West from those jobs and toward international development work. In the remarks I cite earlier in the chapter, Rosemary, Dolly, and Christine describe how their jobs no longer provided them with a sense of personal fulfillment. Susan discusses the lack of professional freedom she had in her teaching job in Britain, and Evelyn's career was nose-diving as funding for educational consultants in Ontario, Canada, evaporated. In their interviews, other women allude to the difficulties of teaching in and administering intercity schools in the UK and to the necessity of gaining more teaching experience to facilitate professional advancement and pay raises back home. Louise, Susan, and Dolly also have the sense that their teaching work in the West is less than worthwhile. The frequency with which my research participants mention work, in all its parameters, speaks to the salience of paid labor to their identities. In this way, they are similar to many working women in the West for whom work is crucial in defining a sense of self (Bielby and Bielby 1988; Gini and Sullivan 1988; Smyth et al. 1999). Job dissatisfactions such as limited control over work situations, relatively low pay, and devalued 'caring' work are also common experiences for many women in the Western work-force (Armstrong and Armstrong 1992; Crompton 1987; Davies 1996; Fisher 1990; McDowell 1997). With the unprecedented entry of large numbers of women into the labor market in the twentieth cen-tury, working women now more often define their social and personal worth in relation to their 'success' at work, meaning that job dissatis-factions and perceptions of failure at work impact on women's sense of identity in more profound ways than in the past (Gini and Sullivan 1988). Yet as gendered divisions of labor continue to restrict the kind of work women do and to define their labor as of lesser cultural value than men's, women often feel 'out of place' in the Western world of work (McDowell 1997).

But even when the gendered, nurturing work of education is responsible for a compromised sense of self-worth at home, my research participants (re)turn to that work as a solution to their esteem problems and as a way to gain more control over and independence in their lives by taking teaching on the road. Teaching abroad empowers women by increasing their knowledge, specialization, and experience, which can translate into professional advancement, work autonomy, and pay increases once they return to the West. But working overseas in educational development also allows them, like their colonial coun-terparts, to realize their intellectual potential, to find more fulfilling work immediately, to travel, and to garner some authority (Freeman 1990; Procida 2002; Ware 1992). Most Western women in Gilgit may still be involved in education, which also remains one of the few nor-matively appropriate professions for Gilgiti women, but the gendered nature of their work is affected by the authority they gain through training mostly male teachers, being Western educated, thinking of themselves—in contrast to Gilgiti women—as independent and smart enough to do the work, and representing that work as an essential cul-tural 'improvement' project. Their self-designation as 'honorary men' invokes a deleterious hierarchical claim to a masculine type of author-ity as much as it describes their understanding of how indigenous peo-ple comprehend and negotiate with Western women. For example, Evelyn thinks "I'm very *male* here. I'm a white woman with a certain level of expertise. I'm given great *respect*, and I have male *patterns* in my life most of the time. I don't have too many troubles, because I'm a man, according to everyone *outside* my family." For those research par-ticipants who feel oppressed as women in Gilgit, experiences of 'male-ness' and authority articulate uneasily with their everyday embodiment in *shalwar kameez* as subjugated 'woman.' These conflicted experiences are never reconciled. However, Western women often employ one contradictory experience to negotiate the other; feelings of masculine authority remind them of their 'real' freedoms, freedoms that seem to fade as their body images change. Despite their ambivalent relationship to authority, coming to Gilgit often helps Western women improve their self-images, as well as their work circumstances, through claims to dominant subject positions.

All but one of my research participants have jobs in Gilgit, and most, like Julia, claim that they would not be in Gilgit, could not exist there, if it were not for satisfying development work: "Work is *so much* your life here, not just part of your life. There's no other reason to be here. And if you're not here working, then there's not really very much else to do. It's quite boring. So that's where people's focus is, in their existence here. That's why you come here, and what people feel they need, to get a certain *achievement* out of their job." My research participants travel to Gilgit because of their work, and the sense of achievement they experience through it helps them develop a sense of themselves as confident, capable, and worthwhile people in various spheres of life. If Margaret "wasn't working, I don't think I'd stay, I'll be quite honest . . . You're donating [*sic*] two years of your life at our particular age, so you don't want to waste valuable work time." Evelyn also understands herself and "this place in terms of work. I don't think I'd be here if I wasn't working . . . It would just be a round of washing the dishes, making sure the kids get to school and that I make something nice for supper. I think, for *me*, that would be *very, very, very* difficult." Rosemary works "because I believe in what I'm doing and I need something to do . . . You *need* to work for yourself, to keep you in action and elevate your mood . . . It's a mental *survival* thing, it's not just financial . . . I also want to contribute, especially with *women's* issues. You can't be rampant about women's issues here. It's got to change in time. But I would work. I would *never* not work." When women whose sense of self has been significantly defined by work in other places find themselves jobless in Gilgit, they can experience a renunciation of self. Jane thinks

one of the hardest things in life is dealing with broken expectations, and I came to Gilgit with the expectation I would work outside the home. And after I came I realized it wasn't practical . . . I *longed* to do something, to be more me, and yet I didn't want to leave my child with anybody . . . So it was a huge adjustment for me to be suddenly at home with very little outside contact and a husband gone. So it was tough the first year or two . . . Now I'm *finally* teaching again, in the mornings.

René is the only research participant who gets little satisfaction from her teaching work: "I feel like the things I do in Gilgit are not things that I do for myself. They're just things that need to be done . . . I just don't have many opportunities here. I teach at the [expatriate] school because they need me, and I want to make sure that my kids have a good education. So I'm there trying to do a very good job for the sake of the kids. But it's not work I do for myself." René would welcome satisfying work opportunities that enriched her sense of self, so she regrets having to make do with the teaching she does on behalf of her children. Moreover, the energy most Western women in Gilgit derive from rewarding work is missing in René's life, which impacts on her ability to cope with the challenges of living in Gilgit over the long term.

These quotations hint at three main ways through which my research participants use work to nurture their identities in Gilgit. First, they understand their work, no matter the field, as primarily a means to gain a sense of themselves as capable, *valuable* women who earn money and make a cultural difference. Evelyn explains this relationship between work and personal esteem most bluntly when she says "If I knew my job here was OK, then I'd feel OK, and if I felt I was being made small at my job, then I would be a *wreck*, because it's obviously quite important to me." Amanda "would like to have more exhibitions and freelance work, because when you sell something you feel much more self-confident and worthwhile." When I asked Andy what the most meaningful aspect of her life in Gilgit was, she, like Janet and Elena, blurted "My job! When my job works, it's the most fantastic job I've ever had, and when it doesn't I've never been so bored in my life. I absolutely love it, the teaching and training. I love going into schools and making a difference. Today I watched a man teach—he must be 55—and he's trying so hard it makes you want to cry." But while many Western women in Gilgit experience development work as personally fulfilling, Joan's sense of self has suffered through it: "I felt particularly important . . . in my high powered job, where I'd been respected by the educational community . . . everyone telling me I was wonderful. And then to come here and *nobody* tell me *at all*, this was hard . . . So it's been a *tremendous* change . . . I'm quite a different person here, obviously, to the one I was when I left England." Various women's sense of self, therefore, has been affected differently through a change in jobs.

Second, my research participants often rely on work as a means to cope with the anxieties and feelings of alienation generated by their circumstances in Gilgit. Work stimulates the mind while time passes. Abbie thinks that "another reason why I can manage here is I get up everyday and go to work. Even if it's for a couple of hours, I need that kind of schedule . . . Otherwise I would be greatly depressed. I'd spend all my time thinking about being somewhere else . . . God wants me here, but I don't really like it. OK, I'm doing a good job and all, and now I'm even finding it fulfilling." When I asked Marion what she did to make herself more comfortable in Gilgit, she referred to plentiful, self-directed work that improves local living conditions, while not stressing her own:

> We've been quite bossy at work, insisting that we *do* have work to do, because boredom is one of the big problems here . . . Work's tremendously important. That fills our time, I suppose, and makes us feel like we're doing something . . . I have certain abilities and skills that . . . it's my duty as a human being to share . . . to make their life better, without making my life worse. I don't find it necessary to make my life miserable in order to share things and make someone else's life better.

Finally, Andy and Marion allude to how work fulfills their philanthropic aspirations for local cultural improvement, as well as their search for personal and professional development. Interweaving benevolent and professional facets of self, Louise thinks that "Professionally, work's the most meaningful thing in life here. I do observations of teachers, and to look back at them, it's lovely to see that they've actually changed because of something I've done." Janet is also satisfied to shape a legacy as a cultural benefactor: "I've begun to think 'Look at what I've done.' And when I looked at that, it's good. This whole crowd of people, who I *know* have benefited from the service I have helped to provide. And there's this [Gilgiti] colleague who's *grown* in stature and assurance and competence, everything. That's not a bad legacy." While Christine downplays the personally satisfying aspects of work, she takes the 'improvement' part of her job seriously: "I pointed out [to the teachers I train] that I was here to do a *job*, and the way we worked in the West was that you took your job *seriously*. I was here to do something, not just to drink tea with them and have a laugh all the time. Obviously I tried to make training as light-hearted as I can, but, at the end of the day, they're to do a job and I'm here to train them to do it better." By stressing her superior leadership, Christine contrasts herself—an independent, competent British woman with a good education and important work responsibilities—to local teachers who are represented as professionally uncommitted, rather than as individuals who work according to a different set of values and a different understanding of an appropriate home/work balance.

To understand how Western women in Gilgit fashion their lives and identities through work, we need to examine the work they do, their impressions about the success of that work, and the effects of work on their sense of self. All but two of my research participants who have jobs in Gilgit are involved with educational (25 women) and health (3 women) development. This 'women's work' includes nursing, equipping and directing women's (mainly reproductive) health clinics, teaching, teacher training, managing an English-language library, and administering schools and educational development programs. Like Rosemary, Fiona, Marion, and Lyn, many of these women sought positions in development projects explicitly focused on improving the lives of indigenous girls and women.

The gendered and mobile nature of much of my research participants' work poses several difficult challenges. Margaret discusses the problem young Western women face when they are confronted by indigenous male teachers, administrators, and village elders who depict them as morally suspect: "Volunteers work really hard in teacher training, but they've heard from some teachers that more conservative, religious people tell them 'You can't trust her. She's a loose woman. She doesn't wear a *chador*, she talks to men. So you shouldn't believe anything she says,' which affects the kind of work she can do . . . This is confusing and frustrating." 'Morally lax' Western women also have to cope with being alone on training courses in neighboring villages, where they are unsupported and alienated for several weeks at a time. NGOs usually provide educators and health workers with very little information about their work context and even less guidance regarding local learning, teaching, and management styles. Women also work long hours, six days a week, to meet the high expectations NGOs set for dramatic development results.

Despite these challenges, Western educators feel driven to 'develop' as many facets of indigenous education as possible, bringing them in line with Western standards and methodologies. Conforming to an ethnocentric logic that serves as a yardstick by which women development workers measure their success, they try to 'improve' teaching and learning styles, teachers' fluency in English, math, and science, the dominant work ethic, classroom discipline, school management and schedules, hygiene standards at schools and in local homes, and the status of indigenous women through educational initiatives to meet a 'superior' Western standard. They also strive to promote educational content that is more "cosmopolitan" in perspective, but still meets the needs of people whose culture, geographical location, and unwritten mother tongues are not addressed in government-issue textbooks.[16]

And most of my research participants believe these changes are valuable and welcomed by indigenous people and, like Jane, are surprised if they learn otherwise: "You have these ideas that you're going to be able to do something and really make a difference to people's lives. And yes, there has been a lot of that, but we hadn't realized that there'd be opposition from *within* communities. Who would guess . . . people are not sure they really want our style of education and have misgivings about what it might do to the structure of society?" Andy thinks that "[t]he methods used for teaching English here are *appalling*, so the children don't know any English and the teachers have no confidence. They have none of the skills they need to teach English. So, we train them how to teach English." While ESL training is certainly more developed in Europe, North America, and Australia than in Pakistan, Andy's comments and ethnocentric judgments about teachers' 'confidence' are also infused with a 'West knows best' attitude similar to Jane's. Ignoring the limited scope of many educational curricula in these locations,[17] Lyn also believes her teaching initiatives can improve local education and indigenous culture by extension: "One of the *biggest* problems is this incredibly *narrow* education they have. Ask somebody where Canada is and they won't have a clue. They have no concept of anything. And the biggest shock to me was that they don't *want* to know . . . So better education comes first, and I think through education that would flow to the culture more generally. We *have* to free up the girls." Using the same racist logic that assumes indigenous people do and know nothing, Rose finds the local work ethic "*unbelievable* . . . It's nice [for teachers] to have this salary, but no one worries about work . . . About maths they know *nothing*. They even can't teach in Urdu . . . I can only show them how to do everything, giving demonstrations. Yeah, and basic things . . . like washing their hands and bringing a notebook . . . These teachers don't know how to do this . . . I hope my example will work." Christine's list of 'civilizing' educational improvements includes

> suggesting [laughs] that they work Monday to Friday, have Saturday
> and Sunday off and no holidays at all . . . You get all these religious hol-
> idays. But I said to them, "On these holidays, do you do anything?"
> "Oh no." And I said, "At Easter and Christmas at home, Christians go
> to church, actually do something religious on that day. But if you don't,
> why not have the holiday on a Saturday?" . . . And their school terms
> are too long . . . At home it's six weeks, then you have a week holiday,
> six weeks, and then two or three weeks of holidays . . . But here, with
> this *endless* term, they can't plan *anything* . . . I think if it was in smaller
> chunks, it would be much better. But you'd need a *revolution* to do

that . . . When we rented accommodation before we built this school there was no toilet. I kept saying "It's *terribly bad* for you not to go to the *toilet*. That's why you've all got kidney problems." . . . I think that's part of education as well. I think using the fields all the time is *terrible* . . . We must get toilets and teach the children how to use them here and at home . . . Let's teach a bit of hygiene . . . Education isn't just about learning A, B, C, it's about many other social things as well.

To summarize, the goal of educational development in Gilgit, according to the Western women who teach there, is personal, cultural, and societal advancement, which dovetails with state interests in replacing local teaching technologies with Western varieties as a global display of national projects of modernization.

By drawing on this sense of themselves as improving philanthropists, which incorporates discourses of gender, race, class, and imperialism, my research participants adopt liberated and authoritative, yet nurturing, identities as they do their jobs (Leidner 1991; McDowell 1997). The caring work of teaching that may have been undervalued at home becomes more fulfilling for Western women in this context as they enlist authoritative discourses of race, class, and imperialism to temper the gendered nature of the work they do. Work thus helps to constitute and confirm their identities as free and superior metropolitan women. And the notion that indigenous women are not allowed to do this work, or do it poorly, substantiates this construction of self. For example, Joan told me that "There are only two or three professional women that I know here who have jobs that bring in money. And they say to their families 'I have a full-time job, I'm having children. I'm prepared to peel potatoes, make the chapatti, but I will *not* maintain animals or go to the fields.' They're creating a different work role for themselves . . . but these are *very* small beginnings." As I have mentioned, my research participants usually depict Gilgiti women as economically dependent house servants with limited spatial mobility and clothing choices. Therefore, they believe indigenous women rarely obtain the education necessary to teach, the opportunity to land a job, and the time to do the work well, despite what they see every day to the contrary. According to Susan,

It's no good saying to a bunch of teachers that they need to mark books everyday . . . If you're a woman, you leave school at half past one, you go home, cook food for your children, husband, and in-laws, you do the washing, you make sure your children are doing their homework. You don't breathe, let alone plan a lesson. I came expecting prepared lessons, professionalism, taking work home . . . but for women it's just not possible . . . In Britain you . . . devote your whole *life* to teaching almost, and then breathe in the holidays . . . You have to change your ideas here to what women can do but don't and what they really physically can't do.

These representations of Gilgiti women's oppression and varied ability to do the demanding work of teaching confirm my research participants' sense of themselves as liberated and always already capable women. They also sustain a certain perception of and sympathy for the constraints some indigenous women, like their Western counterparts, face as workers outside the home, juggling domestic and work responsibilities.

While Western women usually understand their work as a mutually constitutive process of philanthropy and self-development, they are not always convinced that Western initiatives benefit indigenous people in the way development guidelines suggest. Evelyn worries about the monetary and exploitative costs of development: "I am here, making enough money to give my kids a future *because* I'm doing work with kids who have no future. That's a terrible tension [cries] . . . If you look at it in development terms, [fewer expatriates in Gilgit] is very good. I *cost* these people a lot of money, and I gave what I could, but I can't imagine that I gave what they paid me." Fiona perceptively questions the value of imposing Western development agendas, theories, work standards, time-lines, and budgets that are wasteful and harmful to local systems by becoming "aware that there are other ways of seeing and doing things, other than a Western one. The West certainly doesn't have all the answers, neither are their ways the best . . . I think the West is stupid. They see themselves as fairy godmother, supplying all this stuff on what they see as their terms. But when it gets here, the work needs to get done on local terms." And Julia feels foreign development workers deceive themselves by thinking they make much of a difference in Gilgit: "People say that they're sacrificing two years of their life to work here. I can't see what this means . . . There is some *patronizing* idea that they're doing some good, being worthwhile, making a difference. As if I actually have much power to make a difference. It's quite an egotistical, individualistic Western way of thinking . . . If you inspire one or two people, then maybe that's enough." These few women may question the benevolent aspect of their work, but most of my research participants believe they are here to make important cultural changes.

In summary, what are the implications of development work for Western women's subjectivities? First, Western women forge a sense of individuality and personal and professional empowerment through their social philanthropy. Second, and relatedly, my research partici-pants' work serves to define and naturalize discursive differences between Gilgiti women who are ostensibly home-bound, needy, and dull-witted, and foreign women who are free to work abroad as capa-ble benefactors. Discourses of gender, race, class, and imperialism intersect in the construction of these differences to reinforce an array of dominant binaries. Third, identity performances through 'improv-ing' work enact imperial subjectivities and practices. Discourses of imperialism infuse their work as Western women represent Gilgiti peo-ple, in deeply patronizing ways, as unable to improve themselves and development practice as white people saving brown people from self-destruction. By denigrating indigenous education systems, they justify Western cultural intervention, especially into the lives of Gilgiti women, and their 'civilizing' mission. Even when my research partici-pants are critical of some aspects of development and promote less racist and humane interventions into indigenous lives, they generally depict it as an opportunity for doing good in the world, thereby simul-taneously disrupting and recuperating an imperial logic. These domi-nant understandings have a long, but fractured, legacy in South Asia (Jolly 1993; Ramusack 1990; Spivak 1985), as do practices of social, cultural, and intellectual reform through white women's burden of edu-cation and health care (Chaudhuri 2002; Ghose 1998; Jolly 1993; Lessard 2002; Ware 1992; Zoller Booth 2002).[18] While these authori-tative practices and perceptions have always been contested by partic-ular Westerners and thus have shifted over time, they have been further unsettled by antiracist discourses within contemporary feminist and multicultural movements that caution teachers against prejudiced, eth-nocentric worldviews and activities, especially in the classroom. However, some Western women in Gilgit recuperate imperial practices of conversion when they try to mould the lives of Gilgiti women and children to fit a Western prototype and to impose metropolitan, yet ostensibly universally applicable, models, principles, behaviors, knowledge, and values through education—especially in the English language (Jayawardena 1995; Jayaweera 1990; Ramusack 1990).

Free or Not?

Finally, I want to explore how the gender oppression many of my research participants perceive in Gilgit poses a challenge to the con-struct of the liberated, independent Western woman. As they do in most societies, discourses of gender oppression and liberation coexist uneasily in Gilgit. Western women stress their oppression when they feel the weight of vulnerability, anxiety, and frustration. They speak of freedom when they have a sense of control over their lives, such as when they are teaching or talking about their development work, travels, and the status of Muslim women. Even though both perceptions are prominent, experiences of gender oppression are eased through discourses of race, class, and Orientalism that reinforce Western women's sense of autonomy and difference from Gilgiti women. However, these ambivalent experiences of gender freedom and constraint in Gilgit have prompted some of my research participants to reflect on the gendered aspects of their lives in the West.

To recap, most foreign women in Gilgit represent themselves, in contrast to indigenous Muslim women, as free of gender constraints at home. Joan provides a good example:

> One of the reasons I thought it would be nice to come to Pakistan was to get an understanding of an Islamic patriarchal society, and I thought I would be more tolerant of it. But I can't say I have because there's no freedom for any woman . . . Local women have *absolutely* no choice in the direction of their lives. Coming from a Western society, I'm sorry, but that is the thing that we've got to say—we do it well, don't we? I mean, the options you had, the options I had are no different, in a way, from the options of a man, except for the childbearing . . . That's the only thing. All Western societies have got it sussed, haven't they? I know that women fought for it for us, but it's a good situation now, isn't it? They have *not* got it here, and nor is it on the horizon.

But living in Gilgit is another matter. Joan intimates that this "Islamic patriarchal society" oppresses all women who live in it, even liberated Western women. Susan agrees. She finds "it difficult to go into a school with Rick, and *immediately* the head teacher bypasses me and goes straight to him . . . Women are not treated as equals here. I *am* an equal with Rick. There's no doubt about it . . . Teachers should treat us that way, but they don't, and that *really* pisses me off." According to most of my research participants, European, North American, and Australian women forfeit equality, spatial mobility, clothing options, and their uniqueness when they move to Gilgit. For example, Jane "went through a period where I just felt like a tag-along, The Wife. Going to villages, thinking that you're there because you're his wife rather than being enjoyed as an individual. It's about getting recognition and respect as an individual apart from your husband . . . Expat men are treated with more honor, deference, and respect here than they would be at home. Women aren't." My participants also complain that traveling alone is almost unbearable due to their vulnerability. They told me they have to withstand male stares, unwanted sexual advances, Gilgiti men's contempt, and the sense that their bodies are shameful. They almost all agree that respect is not automatically given here, but is won from indigenous people only if Western women are married, have children, act demurely, and prove that they are not morally corrupted. René thinks she has "different relationships on *every* level than a [Western] man would . . . Women don't get the respect that men do. And so then immediately that puts a limitation on me as a woman. And generally local people assume that women aren't as good as men morally, and so a woman has to prove herself, whereas a man doesn't have to prove anything."

Moreover, most of my research participants claim that the oppressive "Islamic patriarchal society" in Gilgit fosters unjust behaviors and attitudes by foreign men, which are presumed to be restrained by gender-equitable Western cultures; these men monitor Western women's behavior and relish their 'new-found' gender privilege. Evelyn objects that "Jim, who, of course, I've seen in a *shalwar kameez once* in over three years . . . is now so full of double standards that it drives me crazy. And [clothing's] one of them . . . He hasn't changed his clothes at all. He looks like he just stepped out of an L. L. Bean catalogue, and he's telling me that I've *got* to wear *shalwar kameez*! Ugh!" Lyn is

> very suspicious of men who like to live in this environment when they don't have to. To me, a Western man who thinks living here is great is immediately suspect. Why do they like it so much? They almost always feel and act more comfortably because it's a man's society, a man's world. Everything's geared towards looking after men. The whole society is aimed at making men feel good, important, and powerful. So, as I said, the men who choose to stay here after their two-year volunteer job is over, I think "Hello, what's going on in your head?"

And according to Amanda, "Normally [without Allan's food and schedule demands in Gilgit] I'm not such a housewife . . . Allan is the big boss in his office, so this has something to do with [the differences in our roles here]. If a woman comes out, she can never be the boss and she has to earn respect. But if you come as a man here, it's easy." Western men even become patriarchs within development institutions where women's attempts to hold top management positions are frustrated. Julia explains that "Within the hierarchy of development organizations, it's absolutely a disadvantage to be a woman. There are even fewer senior women than there were, and it's very hard for a woman to get to management and to survive there. It's a very male establishment. It's very hard to be a woman and be confident enough to survive in this sort of environment." Although many of these gender dynamics are not unique to Gilgit, most of my research participants blame indigenous culture for the disturbing shift in their male compatriots' sense of fair play and role as protectors of Western women.

Western women cope with the gender discriminations enacted by Gilgiti and foreign men in several ways. First, at times they purposefully draw on aspects of femininity to get what they need to survive. Andy

> plays on gender . . . We need to survive, so we all play on everything when we need to. A couple of my girlfriends came to visit, and they needed to get on their flight, but no flights went for five days. They didn't have confirmed seats on the first flight that went out, so they went down to the airport and they cried, and they got on the flight. Half of me says "Great!" and the other half says "That's outrageous. You pushed Pakistanis off the flight by crying." But you can't solve all the world's problems.

"To get people to leave me alone," Evelyn acts "like I'm not competent. But I do have so many other areas in my life while I'm here where I'm seen as the grand Pooh-Bah that it's a great and fabulous relief to play the weak female type they expect." Second, they sometimes rely on Western men who they believe have been unaffected by patriarchal local culture for protection and defense. Third, some women cope by ignoring gender dynamics. For instance, Christine has not "been as aware of this gender thing as I could have been, and I blazoned on in spite of it. I wanted to get something done, so I ignored it. But, again, I think perhaps being older, perhaps it was easier to do that." Finally, my research participants usually draw on discourses of race, class, Orientalism, and imperialism to recuperate a sense of freedom and difference. Their experiences of gender oppression are soothed by feelings of racial, cultural, and moral superiority, as well as through strategies that denigrate indigenous people and institutions while differentiating Western women from Gilgiti ones. These maneuvers thus contest the ability of the Other to manage Western women's lives. Moreover, culturally improving self-images are premised on imperial discourses that provide women with more authority than they usually enjoy in the West.

However, not all experiences of gender discrimination are purified by these mutually constituting discourses. While discourses of race,

Orientalism, and imperialism are used effectively to emasculate Gilgiti men, they do not explain or condone white men's oppressive behaviors. Most women deal with this conundrum, as I outlined above, by blaming a contaminating patriarchal Islamic culture. But a few of my research participants, by reflecting on similarities between Western men's behaviors in Gilgit and in the West, come to realize that they also experience comparable forms of gender oppression at home. Margaret said, "I resent that [Western men don't have to make any compromises here]. I have to change *everything* . . . Being a woman, you've got to make a lot of compromises in your life. I intellectualize it in that sense. And it's forced me to confront those issues here, because living in Australia, I don't recognize those compromises." Amanda uses the generator when electricity is in short supply, "but it's so noisy . . . I was alone, working, and the guard came at 2 o'clock in the morning to tell me to switch it off. He would never come in if Allan was here. This I don't like. The man is *always* the boss. And I can't challenge Allan's bossiness in public or he'll lose face as the boss in the office . . . It makes me think of women's liberation at home. Maybe I've become a feminist!" Evelyn would

> come home from work, and be very tired and have these neck muscles that were like this, and then I'd have to start cooking and remind kids to do homework and make sure that everyone was where they were supposed to be. And [my husband] would walk out the door. Now, I *never* did that to him in Canada. I never said "OK, what are you going to make for supper" when *he* walked in. I would usually have something started. But I never put it on him. It's always on me. Not always, but *often* on me . . . I've come to realize that women in most places are doing double or triple work, and I feel I'm doing double here.

Margaret, Amanda, and Evelyn find that their experiences of gender oppression in Gilgit encourage them to reflect on the gendered meanings of their behaviors in the West and to reevaluate how they are situated and situate themselves in Western fields of power. In doing so, they may identify new avenues for actively intervening in processes of self-constitution.

I say 'may,' taking into account that my research participants seem to construct fairly unified self-representational stories as a way to create meaningful and coherent life histories, to understand how they are embedded in a web of social relations, and to negotiate the variability of life. Western women in Gilgit have enough of a common work, gender, class, racial, and imperial history to tell similar stories about themselves as free, unique, and philanthropic women. Although these narratives are partial—in that they are selective and unfinished—and adaptable, they do seem to limit practices of identity reformation. These stories serve as practical guides to behavior, so subjects are not likely to act in ways that unsettle their self-representations. As McNay (2000, 80) claims, "individuals act in certain ways because it would violate their sense of being to do otherwise." Thus avenues for self-transformation are limited through self-imposed behavioral restrictions. Western women in Gilgit may convince themselves that they are 'free' after all.

Why Did They Come?

In answer to the question I posed at the outset of this chapter, most Western women travel to Gilgit for reasons of self-determination and philanthropy. They have varying recourse to discourses of gender and class in their travel experience to enable them to develop a sense of themselves as independent individuals; women who are more assured, competent, and authoritative than they were before they left home. This process of self-fulfillment largely reinforces the notion that Western women are free. The discursive construct of the 'liberated Western Woman' is perpetuated not only through the experience of travel, but also in relation to the Third World/Muslim Woman who is imagined as devoid of freedom and autonomy. Western women's subjectivities are thus constituted in Gilgit through additional discourses of race and Orientalism pertaining to Other women, discourses that were already prominent during the colonial era in South Asia.

Many of my research participants' decisions about their clothing and the nature and value of the development work they do in Gilgit are infused with these discourses, constructs, and exclusionary politics. They use clothing choices and work to define and naturalize hierarchical differences between themselves and Gilgiti women. Development work also provides them with an opportunity to increase their experience and specialized knowledge—which can affect the likelihood of promotion, raises in pay, and work autonomy in the West—and grants them some authority that tempers the feminized nature of the jobs they perform. Western women's sense of authority, in conjunction with their feelings of moral, racial, and cultural superiority, also helps them negotiate their experiences of gender oppression in Gilgit. Although my research participants usually purify or explain away these experiences, some linger to inspire women to reflect on the gendered nature of women's lives in the West. Whether or not Western women reconstitute their subjectivities through these self-reflections will depend in large part on the coherence and resilience of the stories they tell themselves about their freedom, as well as their openness to the situation of Gilgiti women.

Many Western women are also drawn to Gilgit by their philanthropic aspirations. Most of them come to help, specifically to instigate social and cultural improvements by revamping an allegedly sorry educational system through their greater knowledge and skill. When my research participants represent Gilgiti teachers, especially the women, as incapable of such work, they depict themselves as superior cultural benefactors. Even when a few foreign women question specific development practices and outcomes, and thereby partially disrupt discourses of Western benevolence, many of them adhere to the notion that Western-directed improvement agendas achieve what indigenous people on their own cannot.

Western women usually come to Gilgit hoping to 'civilize' indigenous people and realize independent selves. Ironically, this sense of independence is achieved through the discursive support of Gilgiti women. Without a belief in indigenous women's oppression and lesser capabilities, Western women might find it more difficult to assume they are benevolent and free. My research participants' subjectivities are also formed through their perceptions of and interactions with Gilgiti men. As I described in the previous chapter, discursive relations between Western women and Gilgiti men sometimes shift, producing less rigid boundaries between them. And when those relations change, Western women's subjectivities are also reconstituted. However, similar fluctuations in subjectivity seem not to occur in relation to Gilgiti women. Perhaps the greater and more consistent sociospatial distance between the two groups accounts for the relative stability of Western women's depictions of indigenous women as a subjugated mass. Barbara Bush (2004, 94) argues that in the colonial era "this distance increased the risk of cultural misunderstanding and reinforced white women's sense of superiority." These dynamics seem to be troublingly persistent. As Muslim women remain largely imagined within the frame of an imperial logic, Western women in Gilgit can constitute themselves as free travelers and developers.

Chapter Four

Another Bun in the Oven

The cultural history of imperialism cannot be understood without a theory of domestic space and gender power. (McClintock 1995, 33)

The Cult of Domesticity

To ascertain what systems of rules structure social meaning, particular versions of reality, and processes of subjectivity formation, we need to be attentive to actions and representations that become normalized in specific social settings. In previous chapters I have discussed many of my research participants' customary behaviors in Gilgit, as well as the discursive foundations and (often unintended) effects of those actions. So far I have concentrated almost exclusively on Western women's public sphere activities—working, traveling, shopping, socializing—and their efforts to manage public space. I devote this final chapter to an analysis of how their subjectivities are forged through specific practices and representations within domestic space. However, I avoid reproducing a false dichotomy between 'public' and 'private' spheres. I do this by outlining how imperial, cultural, and racial power relations are constituted and perpetuated through domestic sphere activities, including Western women's governance of their family members and Gilgiti servants, as Anne McClintock suggests in the quotation that opens this chapter.[1]

My nose provided the initial hint about the significance to Western women's identities of making and managing a home in Gilgit. The smell of freshly baked leavened bread cooling on the kitchen counter evoked the cult of domesticity in the first Western home I visited. Following that visit I was struck, in the midst of South Asian fare in the bazaar and Gilgiti homes, by the ubiquity of whole wheat bread in my research participants' diets; thick slices sopped up tomato sauce at pasta dinners, and enveloped peanut butter and jam, tuna, and egg salad in almost every picnic hamper, lunch box, and pot-luck spread I saw. Homemade bread also emerged as a salient and pervasive discussion topic in tape-recorded dialogues. I was initially puzzled by the time and money women spent purchasing ovens, locating yeast, and stabilizing their electricity supply so they could bake loaves of bread when local flatbreads, including *tandoor nan* and *chapatti*, are inexpensive, versatile, and readily available. That they make bread in Gilgit, but are most often content with mediocre store-bought varieties back home, was just as perplexing. Further discussion clarified their actions; when I asked my research participants to describe what they did to make themselves feel more comfortable living in Gilgit, all but one of them mentioned baking bread. For many of these women, baking Western-style loaves of whole wheat bread provides a comforting sense of belonging, as well as a material connection to their lives and histories in the West. For example, Dolly's baking routine satisfies her desire for the solace of 'home': "I like baking bread. You've come into my home and I'm not properly prepared for you. But if I have time I like to make bread, sometimes cake and pastry. I find that such a soothing, homey treat." Kate has "enjoyed cooking and having people over, to feed them a big spaghetti dinner, homemade bread, things which they were really missing, to kind of encourage them. That food makes everyone feel more at home here."

I experienced a similar association when one particularly aromatic batch of bread made me sigh nostalgically as I thought of 'home,' even though baking bread has not been part of my family's culinary repertoire. Notwithstanding the taste, warm flatbread has never conjured for me such a sense of familial and community belonging, which are comforting in a setting of vulnerability, alienation, and exile. These imaginings demonstrate that what is said about bread and home augments the experience of both and promotes their sensory association. Yvonne Tarr (1972, 3–4), in the introduction to my favorite soup cookbook, provides a good example of the prevailing talk about bread in the West: "Bread is love, and warmth, and nourishment, and comfort, and care, and caring, and *taking* care, and time gone by, and time well spent, and things natural and good, and honest toil, and work without thought of reward, and all of those things once had, now lost in a country and a world that has rushed by itself and passed itself, running, and never noticed its loss." I was not only experiencing the smell of bread. That smell was a sensory medium through which the experience of home is formulated. As a gendered ritual of domesticity, putting another loaf in the oven assists my research participants in making a 'home away from home,' recapturing a familiar nurturing identity, and experiencing the imagined solace that supposedly infuses Western domestic space, all important elements of 'stayin' alive' in Gilgit.[2]

Bread is more than a socially meaningful expression of domesticity, or a means to recreate a familiar identity by connecting people to another time and accustomed place. Bread is also one of the primary markers of cultural boundaries and difference (Griswold 1994).[3] What bread connotes in a particular culture—in the West most often it is security, love, self-reliance, and abundance—reflects the values of that culture. And those meanings and values, like cultural attitudes toward women, are often understood by Westerners as a mark of a superior civilization (Griswold 1994). As foreign women in Gilgit bake yeast-raised bread in expensive ovens, they incorporate Western values and modern, affluent, nurturing identities into the recipe. And when they denigrate local breads, they metonymically denigrate local culture. For instance, Amanda claims that "The bread you can get here, it's awful, I hate it. I brought yeast from Islamabad, and a foreign consultant taught me how to make bread . . . But the flour here is also inferior, so we bring imported flour from Islamabad too . . . And then I have to put the generator on for five hours to bake it." Lyn is "going to have a go at bread tomorrow, now that I have an oven. Because the bread here is *crap*, there's no doubt about it. And I like *good* bread, so I'm going to have a little experiment tomorrow and see how it goes." What constitutes 'good' versus 'crap' bread may incorporate taste considerations, although my research participants seldom mention the flavor or visual appeal of one kind of bread over the other. The distinction may also be made on the basis of habit or simply preference. However, the values and identities embodied in the bread, like those cultivated in the larger domestic context, seem to be the most important factor in forming a hierarchy of breads. For my research participants, yeast bread, as a metonym for a more dynamic and materially, technologically, and culturally advanced society, is superior to unleavened local varieties. Bread, then, emblematizes a cult of domesticity and a domestic identity as it marks hierarchical cultural differences within domestic space.

When so many Western women in Gilgit devote domestic time and energy to baking bread, the importance of gendered rituals of domesticity and of the notion of 'home' to their lives and identities abroad cannot be ignored. The significance of this sociospatial site to identity formation returns us to the notion of the spatiality of identity I introduce in chapter two and develop further here.[4] Paradoxically, my research participants struggle to recreate a home and domestic identity for themselves in Gilgit even as they criticize Gilgiti Muslim women for capitulating to a culture that confines them to and defines them through the domestic sphere. The act of making bread—a domestic practice common to almost all Western and indigenous women in Gilgit—draws a cultural connection between women while enabling foreign women to feel somewhat capable and secure in a bewildering social setting. But Western domestic ways of being intersect uneasily with my research participants' self-representations as liberated travelers and workers who, unlike Gilgiti women, are free from the household grind. While experiences of development work and travel may bolster feelings of gender emancipation that are bruised by what Joan calls an "Islamic patriarchal society," their focus on making a comforting home abroad recuperates discourses of femininity sidelined by masculine discourses of travel and paid employment. However, as I will demonstrate, imperial, race, and class relations constituted in domestic space through spatial organization and interactions with Gilgiti servants buttress Western women's sense of gender freedom and cultural power.

This discussion of bread-baking as a sign both of Western women's cultural superiority and of the importance of making and managing a home to their identities in Gilgit provides a preface to the remainder of this chapter, which is divided into two main sections that analyze my research participants' Gilgiti homes as spaces of subjectivity formation and social exclusion. In the first section, I describe how they transform local rented houses into Western homes where they feel comfortable, relaxed, contained, and in control. Discourses and practices of gender, class, race, and imperialism intersect, as they did in the colonial era, to constitute these homes as a haven from an Other world, a space of cultural belonging that incorporates reassuring Western values, ideas, and norms of behavior. The second section examines how race, class, Orientalism, and imperialism coalesce within the home as Western women govern their family members and servants to shape domesticity and perpetuate racial and cultural exclusivity as a component of imperial power relations. Although all of my research participants are concerned with making a comfortable home for themselves in Gilgit, for financial and philosophical reasons VSOs do not usually employ servants. Therefore, my analysis in section two pertains primarily to married women who most often have young children living with them. I include a discussion of how VSOs' different class position as nonemployers provides a discursive site from which to blur racial boundaries and disrupt notions of racial superiority and imperial legitimacy. VSOs develop a critique of salaried expatriates' imperialist practices, such as hiring servants and living materially extravagant lives in a larger setting of poverty, which disrupts dominant imperial relations in domestic space and creates significant tensions among Western women in Gilgit. The cult of domesticity, which denotes both the space of home and the amalgam of social power relations that infuse domestic space, thus involves gendered, classed, racialized, and imperial metamorphoses, in addition to identity forming and boundary managing processes.

Constructing a Haven

The Power of the Cult of Domesticity

As many white women in colonial India did (Saunders 1999), most of my research participants experience living in Gilgit—even when they come by choice—as an imposed exile where the world of good food and friends, enriching leisure activities, comfortable clothes, alcohol, sexually integrated interactions, and 'mod cons' seems to belong to a distant universe. This embodied sense of exile is reinforced through the cult of domesticity, which represents Western homes as safe and familiar spaces that need to be protected from the outside world (see Lasch 1977). Julia describes her Gilgiti home this way: "I have to *gird* myself, armor myself to step out of that gate into the bazaar. But my home is my haven. It is a somewhat Western environment where we can wear what we want, play music, drink, and behave however we want. That somehow is very important, to have that place and space." Examples of this notion of 'home as haven' are illustrated in Susan's discomfort with Gilgiti men visiting her spouse at home, Margaret's representation of her home as the resting place of her "Western soul," Julia's reservations about Gilgiti men attending expatriate parties, and Abbie's reluctance to invite her neighbors behind her garden walls. Foreign women protect domestic space by turning locally owned, rented houses into secluded Western homes that are differentiated from the secluded spaces 'imposed' on Gilgiti women.

This desire to draw a strong boundary separating Western domestic space and indigenous public space signals a desire to secure the Western values, definitions, and territorial identities associated with 'home.' Such values and dispositions are expedient in a situation of uncertainty. As described in chapter two, many of my research participants experience a sense of vulnerability, as well as of exile, in Gilgit, because they fear indigenous men who are perceived to be threateningly close in public space. Consequently, Western women construct the secluded, reassuring territory of home to keep those threatening Others at bay. The boundary of home, by separating Western from indigenous culture, provides them a sense of security and of liberation from the perceived social control of Gilgiti men. It also enables Western women to regulate, in part, the spatial distribution of 'menacing' social groups and to evade some avenues of 'cultural contamination' (Sibley 1995). Thus, by making a secluded Western home in Gilgit, my research participants enact a politics of exclusion (Cresswell 1996). They do so by manifesting a contemporary version of "colonial anxiety" (Stoler 1995, 104), Westerners' racist fears that indigenous cultural elements have the power to change them and 'contaminate' their identity. The space of the Western home frustrates this possibility of 'pollution,' because, as I will demonstrate, the values and gender, racial, and class attitudes cultivated in the domestic domain enable Westerners to live among Gilgitis without 'going native.'

When Western women maintain their identity and self-control through the social barrier of home, their efforts to construct and manage this haven resemble what Foucault describes in *The History of Sexuality* (1978) as the governance of self.[5] The way women imagine themselves and their relationship to indigenous Others in a setting of exile, vulnerability, and alienation prompts them within domestic space to cultivate their difference from those to whom they feel superior and by whom they feel threatened. As I will show, many Western women devise gender, class, and racially coded rules for living and managing themselves, their families, and their homes. The Western civilizing mission thus extends beyond managing Gilgiti people through development work to include an impulse aimed at (re)forming Westerners, at maintaining an appropriate white, feminine, bourgeois sense of self. Drawing on familiar and often subjugating discourses of gender, class, and race, Western women enact an oppositional identity vis-à-vis threatening Gilgitis. As David Harvey (1996, 209) suggests, these discourses of difference simultaneously enact a reactionary and exclusionary politic. In this way, 'home' becomes part of a system of hegemonic social values exercised by those seeking to establish identity hierarchies.

I would argue that the spatial ordering of foreign women's homes in Gilgit also aids an imperial politic by resisting the infiltration of indigenous values and people and by inscribing a set of Western identities, values, and concepts of space and personhood in the social world of Gilgit. The internal organization of 'home' may be part of my research participants' larger struggle to stabilize familiar social meanings by fixing them through materially constructed spaces, but these social values, in clarifying 'otherness' and 'difference,' are also the basis for protecting Western definitions of what is the 'correct' way to be and live. These imaginings of space and personhood make Western homes an arena of imperial social conflict, and demonstrate that spatial orderings are integral to power-laden social processes, including subjectivity formation.

As I have said, 'home' is invoked, especially through the act of baking bread, so it remains largely an imagined ideal for Western women, both in Gilgit and in the West. Most of my research participants attempt with great energy to reproduce it abroad, but they are ultimately unsuccessful due to the lack of commodities available, Gilgit's poor material infrastructure, and the illusiveness of such an imaginary construct. And that lack of success in achieving the ideal of home impacts on the experience of home in the West,[6] as Janet describes: "Well, I've just had a visit home . . . it was quite a painful adjustment to the reality of what that *really* is, rather than what you remember and try to reproduce in the nostalgic haze of when you're here . . . There was real anxiety engendered by what I *thought* would be a *joyful* experience." For my research participants, Gilgiti and Western homes endure as sites of "continual desire and irretrievable loss" (Blunt 1999a, 421). The elusiveness of home suggests that it is not only a figment of a specific cultural imagination, but also an *achievement*, a sociocultural realization of articulating gender, class, race, and imperial relations.

Achieving Home and Self

Sylvia Leith Ross (as quoted in Strobel 1991, 10–11) offers Western women exoticist guidance for maintaining identity and self-control in unsettling colonial space in a description of her situation in Nigeria in 1913: "When you are alone, among thousands of unknown, unpredictable people, dazed by unaccustomed sights and sounds, bemused by strange ways of life and thought, you need to remember who you are, where you come from, what your standards are. A material discipline represents—and aids—a moral discipline." In the social setting of contemporary Gilgit, most Western women cope with an unfamiliar culture through material discipline and day-to-day processes of self-governance that include making a 'proper' home[7] among ostensibly less happy and less well-appointed Gilgiti households.[8] Domesticity, as a feminized discourse, delegates the job of homemaking to women, the angels of the hearth. But femininity is not the only discourse of power at work in the home, as the "moral" implications of "material discipline" suggest. Home, as a unit of *civilization*, is also where racial, class, and imperial subjectivities are constructed.

As Stoler (1995) argues, the making of Western women's middle-class distinctions in (post)colonial space is achieved *in situ* through the racialized notion of civility, rather than being transported from the West. By maintaining a carefully circumscribed, civilized home, Western women nurture moral authority through a "cult of sensibility" (Wilson 2004, 20) and keep 'cultural degeneracy' at bay, thereby sustaining those who occupy the highest rungs of the cultural, racial, and moral hierarchy. A civilized home, constructed through routine domestic practices, is thus a class obligation that denotes middle-class respectability and racial superiority, as it marks cultural membership. Although creating a livable life within the home is not a problematic intention in itself, the effects of that creation are often socially unjust. For example, as a place where racial and class differences are constructed, the bourgeois 'home away from home' becomes another site for perpetuating values, binaries, practices, and hierarchies that support an imperial logic in postcolonial times.

Western women in Gilgit cultivate civility, middle-class respectability, and racial difference in many ways through the "domestic management of imperialism" (Blunt 1999b, 108). Perhaps their most notable effort is appointing their rented houses, especially their kitchens, with comfortable amenities and foodstuffs.[9] Evelyn describes some of these efforts: "You must go to Abbie's and René's homes. They are *posh* . . . If you saw those homes in Canada you'd say 'Wow, this is really nice.' They are *beautifully* decorated, with *borders, hand painted*, going *all* the way around the room. And then in the kitchen you have another border all the way around. I mean, you have to see it to believe it. It's *stunning*." Speaking on her own behalf, Abbie enjoys "the decorating, painting, and buying stuff. I have a sewing machine, but the fabric and tailors here are so cheap, so they do all that stuff. Doing up our kitchen, I spent a lot of time with that, and cooking. This year I have a garden and more vegetables. Jane gave me a bunch of herbs, so I have an herb garden too." Because Esra plans to live in Gilgit for several years, she has "the feeling that it's worth investing in a stove, a fridge, and making bread and cookies. Yeah, I suppose things that make us feel settled and rooted. I also brought over some vegetable suet so I could make mincemeat. The way we live Christmas here is the way my husband's family lives it at home." Janneke's "number one concern was getting a house that we both liked. It's sunny with a great view. We've made it as comfortable as possible with a generator and battery system for lights, running the washing machine, keeping the refrigerator going. For eating, we bring beef and fish from Islamabad, plus some other items we can't get here in Gilgit, like Dutch cheese." Acknowledging her extravagance, Evelyn brings food from Islamabad, from Canada as well, "in embarrassingly large quantities, as in lots of money. We could go to Islamabad and spend Rs. 10,000 [US$165] on food, olives and feta, and cheese, and meat and more cheese."

As Esra mentions, homemaking makes most Western women in Gilgit feel settled, rooted, comforted in an unfamiliar social milieu. Consequently, Abbie's questionable advice to newly arrived foreigners about settling in is to "spend *a lot* of money. I don't know if that's good from a local standpoint, but buy a car, a generator, and get a house with good water, geysers, and a big kerosene stove like we have at the school. We have one of those in our living room, and then we have some small kerosene space heaters that we move around. Oh, and then this last year we got a gas heater for our bedroom." Without Abbie's ambivalent sensitivity to recently circulating development, human rights, and anti-imperialist discourses that dissuade wealthy Westerners from displaying colonial-style affluence in a local setting of poverty, Jane agrees that

> working on the house was how I came to feel more comfortable here. This is our first proper home, although it's rented . . . I feel it's *ours*, so I've enjoyed getting Ikea curtain fabric in Islamabad and stuff like that. Now I look at the house and see the way it's ended up, it's very different than what I expected. I wanted much more of a local house. And look what I have! I didn't want it to look like some sort of Western thing that had been beamed down into Gilgit. But we've ended up living in a much more Western style than I had intended . . . Probably without realizing it, you end up with a dining table, a sofa, battery light system, hot water, and there you are. We're living in America again [laughs].

For Lyn, "Music, books, food is important to feel at home . . . I do think it's important to have somewhere that you feel comfortable and feel is home. Home is where we are alone together, and now that we've fixed [our home] up it's quite comfortable and suits us. Now we have power and water and hot water. It's luxurious."

Many of my research participants imagine that their cultural refinement is partially reflected in these amenities—from expensive battery light systems to imported food to hand-stenciled borders—that distinguish Western homes from most Gilgiti households, which are frequently depicted as bereft of comfort and delicious food. Generator-powered stoves and the loaves of bread Western women bake in them are thus signs of civility, as is the laying out of a garden in this hot, dry landscape. The desire to harvest a private crop of vegetables and herbs in this desert environment, when the bazaar is brimming with them, is allegorically linked to Western bourgeois notions of order, civility, and productivity.[10] Controlling the fecundity of the household yard is akin to taming the environment on metropolitan terms. Gardening is also culturally competitive: foreigners feel they cannot be outdone by Gilgiti farmers, who, as experts in irrigation agriculture, coax substantial harvests from the desert year after year.

Most Western women are equally preoccupied with keeping their homes clean. They are often disgusted by the dust that blows into their houses every day, as were many middle-class white women in colonial India (Bush 2004; Fleming 1992; Hall 2004; Procida 2002). Overgeneralizing, Amanda "never thought that it could be so dirty here . . . I don't know why local people don't clean. I'm so upset about all the dust and these flies. I often wonder how these people can survive." Andy never knows "what dirt's going to make you sick, but you can try to minimize the risk. And you spend your whole life doing it here. When I cook for the [Western] children, I'm just paranoid about poisoning them. Honestly! I'm *so* careful with the food." Elena deals with the potential ill effects of dirt by

> boiling water for 20 minutes and then filtering it. We wash our vegetables well. We don't wash them in iodine though, that makes them taste bad. But we wash them in filtered water. So we take very good care at home. We're fanatical about it really . . . We went to visit one of our teachers in the hospital when she had a miscarriage. It was terrible, it was. The wards were filthy, and there was a cat lying on her bed! It was *filthy* and disgusting.

This widespread aversion to dirt leads many of my research participants to be fixated on expunging all traces of the dust that constantly swirls through Gilgiti streets and neighborhoods. Even though Evelyn, as a notable exception, characterizes herself as the "worst housekeeper I've ever met who can't clean anything," Lyn is convinced that "Western hygiene and living standards are quite different, much higher . . . Most accommodation outside our home is filthy, and that's a real problem for us . . . When you're on holiday, you can put up with almost anything. But when you actually have to live and work and spend three weeks in a dirty, grotty hole, then it's a huge problem." Sustaining the clean Self/dirty Other dichotomy, Margaret agrees that "Westerners have a different notion of hygiene [than indigenous people]. I mean, you have to create your own home living environment which you can identify with, be happy, comfortable and *clean* in. So we've bought some kitchen tables and high counters for chopping food, and we can have as many washes as we like." The association of identity and comfort with a clean home is echoed by Marion: "Well, first of all we cleaned our side of the house from top to bottom to feel better. That made it easier to live here . . . I like cooking, but not sharing the kitchen. I find that probably one of the most difficult things to cope with, because I understand myself as having a certain level of cleanliness . . . when kitchens are communal, the highest level of cleanliness is what we should go for." Perpetuating dominant imperial and class relations from the colonial era, these women imply that a properly clean household reflects a civilized society, and within that implication lurks a bourgeois morality that equates cleanliness with cultural godliness.[11]

When I asked Janneke how her life would change if she had to live like her Gilgiti neighbors, she, like Amanda, Margaret, and Lyn, associated a change in ways of living with a change in rituals of hygiene, with a turn toward domestic slovenliness:

> It would be hard to change my Western ways if I had to keep house like the locals do. I mean, they don't. With cobwebs all over, I don't think they *ever* clean their houses. And they spit and throw water on the floor. I couldn't live like that unless I could bring along my Western ways of cleanliness . . . Sanitation hardly exists here. We've gone into houses where they just have a dark room where everyone just goes in and does their routine. When we inquired how often they cleaned out the "toilet room," we were told, "Oh, once in a while." It's filthy and they *know* it's filthy, but they just don't clean it.

Once again, my research participants position 'dirty Gilgiti culture' in a binary with 'clean Western culture.' This subordinating affiliation between dirtiness and Otherness is so strong that dirt, like bread, takes on metonymical power; it symbolizes abject Others whose bodies and culture are potentially contaminating to Westerners. Western bourgeois rituals of domestic hygiene metaphorically purify and preserve white bodies from contamination in the cultural contact zone (Bush 2004; McClintock 1995). Western women's home cleaning practices thus constitute a form of imperial control that frustrates the possibility of moral, class, and racial degeneration by demarcating boundaries between racialized bodies and hierarchically arranged cultural groups. Moreover, the 'whiteness' that results from cleansing rituals racializes Western domestic spaces, and marks these homes as sites of cultural and moral superiority and as demarcated spaces of civilization within the larger cultural territory.

Many Western women may be threatened by the cultural and racial connotations of Gilgiti dirt, but they can be equally unnerved by a shortage of domestic privacy. Christine "wasn't prepared for, and I don't think you can ever be prepared for, this different culture's lack of privacy. That is something which, no matter how hard you try to tell somebody about it, until you experience it you can't understand the discomfort of it." Privacy, like cleaning and home renovating, is a way Western women can cultivate racial and cultural distance, although in this case detachment is more spatial than metaphorical. My research participants, especially those who live in buildings attached to a Gilgiti household, want their homes to be sequestered private spaces, respites from omnipresent indigenous people. According to Fiona, "I wouldn't move in with somebody I worked with. Even when I say I miss having people to talk to, I like my own privacy . . . I definitely wouldn't want to be sharing a house with somebody I didn't know very well or that I wasn't compatible with . . . I need my own place. I definitely wouldn't like communal-type living or sharing with somebody, especially a local person." Louise admits that

> living within a family compound has its ups and downs. Sometimes it's *very*, very noisy, and I have a big problem with lack of privacy . . . You know, you've just come home from work, and you want to sit and have a cup of tea, or just be on your own, and the kids come in to have an English lesson or to just play or talk. And I feel like saying "Get out! I've had enough today! I need a break from you lot." And sometimes I do that, obviously. I'm not that silly, but I can't always, because I feel a little bit awkward about that. I can't say "Enough. I'm paying rent, and I need some privacy."

Rose is "*very welcome* in with the family, but I'm too much a Westerner to stay the whole day around the stove being warm on the front and cold on the backside, with eight shouting children around you, grandmother who is more ill than well. *You miss your privacy.* So at the end of December, I had to leave." Rosemary has recently moved out of a family compound and into her own place: "Now I *never* feel that I'm under scrutiny. Sometimes the family that owns the house is on the roof doing something or they're working on the terrace next door or they're walking past my windows, but it doesn't feel like scrutiny anymore. They *don't* come in . . . It's a private haven up there. It's bliss." Those of my research participants who desire privacy reveal their need for a physical boundary that establishes space where they can be kept away from ever-present Others. This need to organize domestic space is interrelated with their desire figuratively to 'own' property in Gilgit, as many of them do in middle-class neighborhoods in metropolitan cities. In addition to physical space, privacy provides my research participants the emotional space to cultivate the moral courage they need to live among Gilgiti people.

To respect the privacy of others, as well as to have it yourself, is a further sign of civility and class respectability for many Western women in Gilgit. According to Christine,

> To begin with, when I didn't lock the door, local folks would just *come in*, they'd sit down and just stare at me . . . I mean, *I* wouldn't even have the indecency to walk into my *daughter's* house, unless she'd given me the key because she was going to be out. I would ring the bell and I would wait for her to open the door. They can't understand that here at all. I mean, they're not really being mean, it's just their *different* way . . . They have *no privacy*. There are 23 of them living in that room over there. They eat, they sleep, they live, they die, they do *everything* in that room together. They've never had a secret from anybody in their lives. They haven't got any proper sort of *space* . . . and the more children they continue having the less space they're going to have.

Jean supports Christine's characterization of indigenous people as lacking civility and decency by comparing them to refined, urbane Mongolians: "It will be a lot easier to be my Australian self in Mongolia, my next placement, because they're a sophisticated, multicultural, urban population that is accepting of other cultures and other ways of being. So, I'll have much more privacy there. Then you can be yourself, especially with your partner. That instantly reaffirms who you are." Notions of civility merge with a bourgeois sensibility in the perception that culturally sophisticated people ignore what goes on behind closed doors, knock before entering a room, plan family size to allot an appropriate amount of home space to everyone, and distribute domestic space so that family members, especially parents and children, are kept a decent distance apart.[12] Class, sexuality, and race intersect here, where civility encompasses sexual control as much as cultural decorum.

Hospitality is yet another facet of Western women's domestic management of imperialism in Gilgit. In Evelyn's description of an expatriate anniversary party, we see how white bourgeois civility, and thus racial difference, is cultivated through being a gracious hostess, once personified on North American cable television by Martha Stewart:

> We get invited to lots of dinners. The other night Jane had a dinner party that was *exquisite*. It was for Abbie's fifteenth wedding anniversary. She

had smoked trout. Her husband fishes, and he has a smoke house, and it's like eating smoked salmon. It was absolutely sophisticated and *gorgeous*. So we had fish, Moroccan lamb, strawberry coulis on a cheesecake. *It was just unbelievable.* And it was outside, and she had hung decorated kerosene lamps in the garden. It looked like something out of Martha Stewart. It was lovely, absolutely amazing.

As it was in colonial times (Chaudhuri 1988; Hall 2004; MacMillan 1988; Procida 2002), being a gracious hostess to friends and visitors is a vital component of civilized Western homemaking in the contact zone and of bourgeois women's domestic identity formation. René, for instance, understands herself as "a very relational person, always creating social situations, hosting different parties, initiating dinners. I have a real interest in hospitality. I mean, getting friends together, cooking great food, and being able to have good conversations and fun and, you know, pulling people out of their routine. It's just something I enjoy doing." Evelyn extends that hospitality to foreign visitors to Gilgit, as well as to friends: "We have dinners with our friends in town, we invite people over or they invite us. But then there's a lot of times when people are in town for a short period of time, we see them a few times, invite them over for a nice meal, and then you may never see them again or you might see them only the next year for another dinner."

Playing the gracious hostess and attending dinner parties, therefore, constitute additional forms of class identity in the home for these married Western women. Moreover, hostessing is important to the domestic management of imperialism as a way to represent Western culture in an alien land. The efforts Abbie makes to recreate a Western Christmas celebration in Gilgit is a good example of this kind of cultural protectionism. She explains that

> Christmas is a bit pitiful here, but we try. We find some little branch somewhere for a tree, we invite all our friends and have a meal together. I've usually saved presents that I've brought from home and have hidden for months. You can buy a turkey in Islamabad, so we've done that in the past. And, of course, you can have your potatoes and homemade bread and traditional things. It's good. It's more quiet than home, but we sing Christmas carols anyway.

Roasting a turkey, decorating a tree, and singing carols, like hosting dinners, are social conventions that constitute civilized behavior for Christian-influenced Westerners in Gilgit. Most women in this group use the word 'hospitality' to evoke these refined performances.

According to Jane, for example, "I've discovered that I enjoy cooking. I've always enjoyed it, but even more here, making good food, having other foreigners over to nice dinners . . . I don't think I'll forget the importance of having an open home, of being the person to say 'Oh, why don't you stay for supper.' We can't lose that sense of hospitality and community." Other than Saturday night church dinners, Andy enjoys "organized parties. Of course there are enough VSOs for lots of birthdays to come around quite regularly. And then I invite people over a lot. *Lots* of times I have people for dinner here. That's a great kind of hospitality." Offering Western forms of hospitality, hostesses protect Western cultural traditions in Pakistan and sustain cultural exclusivity.[13]

Finally, most married Western women who are long-term residents in Gilgit labor to provide their families with a safe, healthy, and happy home environment—the Victorian domestic archetype—as part of their efforts at civility. Although Evelyn disrupts this archetype by depicting her home in Gilgit as yet another site of work and relationship anxiety, Janneke describes her homemaking endeavors as familial support measures: "It's important that I give support to my husband in his job by being at home for him—not only as his wife, but also as his friend and companion—and making a happy home, because the social life is so impoverished here . . . If you don't have a happy home in a place like this, it's then all work for him." Jane is most concerned about fashioning her home into a happy, ordered sanctuary for her son:

> I longed to work, but I didn't want to leave my child with anybody. He was cruising around, getting into trouble, needing some discipline and structure, and I felt that he wouldn't get that discipline and structure I wanted him to grow up with unless I was there . . . I also realized that my husband was working a very busy schedule, six days a week. So it was a huge adjustment to be homemaking all the time, but it's made everyone happier . . . I also realized that these are my son's formative years, and I wanted to affect his view of "home." We've made a *wonderful* environment here for a child, with all this space.

Not surprisingly, Western women's self-representations as happy homemakers contrast sharply with their portrayals of 'dispirited' Gilgiti households. Rose tells a story about her host family's spousal relationship: "After dinner, Gulam's wife sat on the men's side of the room for some reason, and he started to touch her. She *really had a reaction*, being afraid of him. She went quickly to the women's side. So strange. I was impressed by the look in her eyes, it wasn't happy at all, just *angry* to be the breeding partner of the man and doing all this house work . . . She's right, being treated like this." Although Julia finds "that being in a large family here is just very nourishing," Marion thinks "this extended family network thing is a problem. I hate to say it because it's such a cliché, but mothers-in-law are always a huge problem for local women . . . They have to tie in with what their mothers-in-law want and her way of doing things, which leads to a lot of domestic strife." Rosemary overgeneralizes and exaggerates to argue that "People go to the doctor just for the attention, particularly women. They get *no* attention. There's no touching or caring in the family. They're just *things* . . . So all the things we know people need, like touching, this culture denies them. I reckon that's pretty tough."

As these remarks show, my research participants commonly depict Gilgiti cultural practices and attitudes associated with domestic life— including extended family networks, spousal interactions, and rules about physical touching—as socially, physically, and emotionally destructive. In this cultural environment, they think, using an imperial frame of reference, that a comfortable, happy Western home is the primary front on which to battle this potentially 'contaminating degeneracy.' René is convinced, for instance, "that it is worth time, money, and energy to try to make yourself a comfortable home so that emotionally you don't get drawn down and you can manage life in Gilgit." For Abbie, "Gilgit and work is at the hospital and not in my house, and so when we go home we're by ourselves. It's a haven. And, you know, over the years we just keep doing more and more stuff to our house to make it a more comfortable and cheerful space." Making a happy home that wards off a regressive culture is a class-appropriate performance for this group of Western women. This social activity also maintains dominant gender, racial, and imperial relations by preserving Western homes as civilized spaces in a larger cultural wilderness.

In sum, the efforts undertaken by Western women in Gilgit to transform local rented houses into Western homes integrate discourses of gender, class, race, and imperialism. Creating a haven from an Other world through the cult of domesticity involves making 'home' into an exclusionary space of cultural/racial belonging that incorporates reassuring bourgeois values, ideas, and norms of behavior. Western homes in Gilgit are thus power-infused sites of the mutually constitutive processes of identity formation, sociospatial organization, and boundary management. By means of gracious social rituals and rigid cleaning etiquette in the home, Western women maintain a white, feminine, bourgeois sense of self, which contrasts sharply with most of their representations of local domestic subjectivities. This contrast is sustained even when Christian missionary women occasionally recognize some solidarity with Gilgiti women in their common domestic duty. The continuous attempt to establish social difference through oppositional identities further ensconces racial and class boundaries. Imperial power relations are inscribed in the space of home not only as my research participants etch Western cultural traditions, values, and concepts of space and identity on the wider social landscape in hierarchical fashion, but also as they conceive of their homes as sites of cultural/moral superiority in this cross-cultural setting.

Governing the Haven

Managing Servants

At this point I want to expand the concept of the domestic governance of self[14] by discussing how the civilizing mission of married women with families in Gilgit extends beyond self-discipline to include the noncoercive management of Gilgiti servants and Western family members within the home. I begin by outlining the interwoven processes involved in supervising servants, and then turn my attention to related dynamics at play in the daily reproduction of expatriate children's lives.

After their interactions with the Gilgiti people they meet through work and other foreigners, my research participants associate most frequently and intimately with domestic servants whom they or their landlords employ. *Chowkidars*—men who guard domestic property and maintain the exterior of the home—are probably the most common servant they encounter. René reveals some dissatisfaction with her *chowkidars'* work when she told me she employs "a man that comes part-time when I'm at school to clean the house. But we also have a day *chowkidar* who's also *supposed* to be the gardener, and then we have a night *chowkidar*. He's also *supposed* to be the mechanical fix-it man. So we actually have two and a half people." As René describes, indoor servants may not be quite as common as *chowkidars*, but the heat, dirt, and work involved in cooking and keeping a household in a setting where supplies of electricity, cooking gas, and water are unpredictable means that these servants are often indispensable to sustaining Western families in Gilgit. Although Andy does not mention servants, she outlines precisely why they are invaluable to Western women with families:

> I couldn't be a married woman or mother here, not for a prolonged period of time . . . So many things are physically challenging that if you

add children to that dynamic, then I think that's difficult. Now, most of the women with children have extra stuff, like vehicles, electricity, ovens, refrigerators, and generators. But I still can't imagine being able to generate enough food for a family here. I love cooking, and it's quite easy cooking for just me when the freezer bit of my refrigerator is working. But if you're boiling kidney beans every night for a family of six, it's just not funny. Or baking your own bread, and laundry, and ironing, and sweeping out bedrooms, and trying to control the amount of dust that comes into the house every day.

Susan explains how servants are also a crucial component of volunteer life: "We're very lucky to have our landlord's *chowkidar* who changes gas cylinders, gets kerosene, goes to the post office, goes to the bank. We're at work, so we can't do these things . . . He will bend over backwards to make sure that our lives are comfortable." Even when VSOs, who do not employ servants,[15] have access to *chowkidar* labor, they spend much of their nonworking energy in reproductive activities. But servants are, and have traditionally been, perceived as a required luxury for married Western women in (post)colonial settings—an indulgence most of them cannot afford at home[16]—because they are often unable to cope with the countless hours of domestic work alone. Western women, who operate in the home in close physical proximity to servants, thus rely on Gilgiti labor (as opposed to that of other family members, especially their husbands'). However, this dependence and propinquity seldom guarantees that they will overcome their perceptions of Gilgiti people to narrow the social chasm that separates foreigners and locals.[17] Rather, this group of research participants usually incorporates servants into the larger processes of racial exclusivity I have outlined above.

Based on the racial and class asymmetries that sustain white employers' ability to issue demands to Gilgiti servants, these Western women deploy Orientalist representations of Gilgiti servants' lives to create and perpetuate absolute difference. They do so even when they seldom witness servants in their own surroundings or consider that domestics have reasons for concealing their thoughts and abilities from foreign employers as part of their 'hidden transcripts' of resistance (Scott 1990). Invoking Marx's (1976, 125–154) labor theory of value, servants may be required for their use-value as domestic laborers, but their exchange-value is realized in their function as boundary markers who distinguish white from not-white, order from disorder, private sphere from public sphere, and civility from cultural decay. In this way, the work of affirming white middle-class identities in Western homes is relegated in important ways to poor Gilgiti domestics

(Bush 2004; Chaudhuri 1994, 2000; McClintock 1995; Stoler 1995). In addition to their function as boundary markers, male servants are disciplined by Western women who, through supervising their labors and cultural improvement, recuperate feelings of gender emancipation threatened in a wider setting of perceived vulnerability and oppression. This asymmetrical relationship in the domestic sphere contributes to Western women's imperial subjectivities, (re)producing imperial power relations on a household scale.

Whether in colonial Africa, Indonesia, or South Asia, indigenous people who worked as servants for Western colonizers were variously described by their employers, using an ethnocentric rhetoric, as troublesome, unpredictable, primitive, inefficient, stupid, childlike, dirty, irresponsible, lazy, and alien (Barr 1976; Chaudhuri 1994; Locher-Scholten 1998; MacMillan 1988; Paxton 1990; Procida 2002; Stoler 1995; Tranberg Hansen 1989). My research participants commonly (re)circulate these degrading representations to detail their servants' behaviors. For example, Jane describes her maidservant as inherently dim and untrustworthy:

> I did have a lady work with me in the mornings when I was at home, doing the cooking and ironing. But she started stealing from us. The temptation was to take stuff whenever my back was turned . . . I was irritated, but you have to try to understand where they're coming from, how they perceive things. They're not able to view it from our perspective, because their experiences are narrower and they haven't had to look at things from different angles. Education enables that, and she hasn't had much education, combined with a very small life. I don't mean that negatively, but she has not had much experience outside Gilgit, she hasn't seen much of the world. So she can't see things from my point of view, or understand my lifestyle. So I've had to try to understand hers.

In a similarly patronizing vein, and without considering his motives or 'hidden transcripts' (Scott 1990; see also Procida 2002, 101), René depicts her gardener as "incapable of thinking about cause and effect. He doesn't understand that if you tie a sheep up and its rope is 15 feet long and there are flowers planted 10 feet away, then the rope is going to allow the sheep to eat the flowers. He doesn't think that *clearly*, you see, he hasn't learned that critical thinking skill." Besides the "day and night *chowkidar* and the house boy that comes during the daytime to do all the cleaning and make the breakfast and lunch," Janneke employs a

cook who prepares the evening meal. I laugh when I think of our poor cook. We had to teach him to "unlearn" using spices because I can't eat curry . . . I put a ban on chilies in the house. *Very gradually* he learned the house rules . . . My experiences with my house personnel have made me see the mentality of the Muslim culture, which is quite different than my own. Every action, like cooking, is done thoughtlessly, in the name of Allah and tradition.

Amanda sustains this portrait of supposedly inept and childlike Gilgiti servants: "I still have a cleaning lady, but I'm disturbed if she is around me, always asking questions. I'm not really her *mother*. So she only comes a few times a week. Such a difference in how she cleans, it's disturbing! So I have to always tell her everything."

Evelyn offers some strong opposition to these hegemonic, subjugating representations of Gilgiti servants by reversing the disciplinary gaze and contemplating the minutiae and global context of her servants' lives. In an effort to understand how her servants view her as a Westerner and an employer, Evelyn explains that she is not

> one of these people who can keep a million balls in the air. One ball and I start moaning. So, this is a clear picture to the people who work for me. I have somebody to clean my house, somebody taking care of my kid until 4:00 pm, there's a gardener outside, and I'm still moaning around them. I don't really have their skills and experience, and I don't really have a lot of tolerance for huge obligations here. So I've learned in a context where people are taking care of me that I can emotionally afford to be taken care *of*. What will they think of me?

With her son's ayah[18] in mind, she recognizes how

> *extraordinarily* privileged I am to know there is food in the cupboard, I have Rs. 1000 [US$16.50] in my wallet, and if my kid gets sick he gets a helicopter lift out of here. My neighbor's six-month-old boy got sick. He died. And she was at my house working. She takes care of my son, even when he was very ill. She *bathed* him the morning we took him out to the hospital. I know how life works. My child *should* have died, and he's alive. Her child had pneumonia. He should still be here, but he's gone . . . I feel *sick* when I talk about things like that. I don't deserve this luck, it's an accident of birth that I was born in Canada . . . But so many people here have so *much* asked of them, and really so *little* has been asked of me . . . They were born in the wrong place and got invaded by the wrong people, and it all got screwed from the very beginning. Colonialism has 99 percent of what is happening here to answer for.

Evelyn's remarks position Westerners in Gilgit as colonial beneficiaries and grant servants individuality, humanity, and competence, thereby providing some new discursive terrain upon which to build less hierarchical relationships with Gilgiti people.

Despite this important disruption, Western women's overwhelmingly Orientalist representations of their Gilgiti servants serve to flatten out locals, to reduce this variegated group of indigenous people to an indistinguishable mass of incompetents. My research participants' ability to homogenize their domestics is founded on their "positional superiority" (Said 1978, 7), on the relative upper hand they wield in situations where indigenous domestic labor remains largely invisible, economic means are inequitably distributed, and discursive authority is uneven. The abiding issue is the process of making difference, specifically the hierarchical 'us' versus 'them' division based on class and racial perceptions. Western women frequently draw on Orientalist discourses to constitute servants as immutably different, as a class whose thought processes, needs, and problems are radically unlike those of their employers.

Three salient practices of homemaking are infused with these social constructions of difference. First, by equating Gilgiti servants with mentally inferior children, my research participants legitimate and provide the moral justification for their imperial practices of employee surveillance, including "policies of tutelage, discipline, and specific maternalistic strategies of custodial control" (Stoler 1995, 150). According to Western employers, if 'immature' Others do not have the wherewithal to comprehend a foreigner's way of life, then they must be kept under constant supervision and taught how to cook, serve, and dress in a fashion that befits a Western home. René, for instance, gets "frustrated with local people whom we're trying to get to do a particular job in our home. They just don't understand, or they don't care, I don't know what it is . . . I get frustrated, but I have to *micromanage* so much because these people seem incapable of independently doing something."[19] Amanda, in contrasting her sense of agency and resourcefulness with what she sees as Safina's lack of it, is disheartened by "the cleaning lady" who is "always around with the suffering face. The washing powder box was empty, so she sat there on the steps of the sleeping room, the box hanging from her hand, with the face suffering. And I asked her, 'What's the problem? I'll give you money and you can buy some more, or I'll send the *chowkidar*.' I have to guide everything for her." Speaking about the man she has hired to clean her floors, Marion complains that "unless we are here to supervise him, he really doesn't know how to do it. He doesn't have any idea that when you wipe things down with a damp cloth you don't do it with a filthy, filthy, filthy dirty cloth. So as long as someone is here to oversee him, he's fine. But he never seems to learn to do it by himself." By fabricating a group of radically different Others who require direct and constant guidance in their domestic duties, Western women justify their imperial discipline of indigenous servants as they attempt to teach them Western ways.

Second, regarding tutelage, Rose has tried to prevent her host family's *chowkidar* and his family from "breaking their teeth on these improperly stored dried apricots. I soaked them one night, cooked them for the family. In five minutes it was finished, they liked it so much. But they didn't copy me, but continued to eat the dirty, dry apricots. I made raisin bread. In five minutes it was finished too. And then I told them how to do this. They have all the stuff, but they wouldn't do it." Here Rose exemplifies how, as maternal figures, Western women often try to improve servants' lives based on representations of difference. The naturalized dichotomy 'primitive indigenous laborers/civilized white mothers' makes Western women feel responsible for their servants' protection, health care, and social uplift. Although Margaret does not employ a servant, her landlord's gardener grows food for her, and thus provides her a chance to enhance indigenous diets: "We brought back good seeds from England . . . The gardener mixed the seeds in with their chard. But he fed half of it to the goats . . . We took some beet root to him, to improve his family's diet, but he had one bite and made a face. It wasn't that he really didn't like it, it's just that it was a new food, and he wasn't prepared to allow himself to improve his diet." Amanda hopes to teach her maidservant how to be more 'human' by providing her with proper health care: "The cleaning lady, she is always sick. This past time it was a stomach problem. Now she brought us a big bill about abortion. I don't know if it's a miscarriage or an abortion . . . We always pay the bills when the servants are sick, to show what it is to be humane, but I don't know why they are sick so often." Similarly, as René depicts her *chowkidars* as "wretched" people with "very sad lives they can do nothing about," she focuses "on small ways to help so I don't get overwhelmed. You try to help in every situation that arises, like showing them how to be generous by giving them time off when they're sick and paying their medical bills. It's an emergency, it's unplanned, and they don't have the funds for it. So we just have to be a generous example." By teaching servants through 'example' how to think, work, and interact socially in a principled, 'logical' fashion, my research participants hope to spread Western culture, including 'appropriate' values, from their homes to indigenous households; they anticipate that their employees will take that knowledge back to their private lives to improve indigenous culture, just as they do in their development workplaces.[20]

This diffusion of Western culture through the labor process serves to bolster imperial power. By disciplining servants to keep them under control and to convert them into a metropolitan prototype, Western women propagate an imperial logic in domestic miniature. Although the ideal is seldom realized by Westerners no matter where they live, most Western employers understand their Gilgiti homes as models of care and equitable spousal relationships for their domestics, as places where husbands and wives supposedly eat, socialize, work, and raise a family together (Chaudhuri 2000). In so doing, they present a nuclear, monogamous family structure as the 'ideal' organization of home. Following an ethnocentric logic that grants them the license to be judgmental, these women hope to transplant this ideal into Gilgiti households via their servants to free Gilgiti women from purdah, 'base' conceptions of love,[21] and 'oppressive' extended families that include threatening mothers-in-law and, on occasion, second wives.

Third, Western women's representations of Gilgiti domestics as an indistinguishable mass of children impede most attempts to see servants as people with rich lives and specific desires of their own. The amorphousness of indigenous servants is so pronounced for a couple of my research participants that they never ask their servants' names, or they refer to servants using imposed European names. For example, even though two years have passed since Amanda hired a "cleaning lady," she still has not asked the maidservant's name, let alone where she lives or how many children she has. And Esra, who hires a Gilgiti restauranteur on occasion to cook, serve, and tidy up at Western parties held in her garden, calls him 'Oscar' as an European corollary to 'Asgar.' As Edward Said (1978) has observed, the power to individuate, or not, is associated with the power to name. Naming is infused with power, because ignoring Gilgiti servants' names or conferring European names and associations on them is tantamount to denying them a self-designated identity and individuality and keeping at bay intimacies and responsibilities they do not want.

As Western women adopt Orientalist discourses to treat their servants as incompetent inferiors, they deny their domestics' individuality, justify employee surveillance, and fortify imperial power. Moreover, Gilgiti servants are symbolically neutered when they are depicted by their employers as children, making them less threatening to Western women. Yet despite the unjust effects of these domestic practices of power, disruptions to dominant social relations in the home are also enabled through this labor process. First, contemporary civilizing efforts within the home are predicated on my research participants' assumption that Gilgiti servants can respond to cultural improvement initiatives to become more cultivated than they were before metropolitan contact.[22] In other words, for Western women's domestic mission to work, servants must be seen to have a mutable subjectivity, if not a sovereign individuality, which poses a challenge to discourses of Orientalism and imperialism. Second, when my research participants play mother to their servants as middle-class tutors of 'generosity' and 'humaneness,' they incorporate a class-based superiority with a race-based authority. In this instance, they may sometimes perceive domestics more as a different class than as a different race of people. This discursive blending can unsettle the racial foundation of imperial practices by disrupting its justification; if servants are not always already inferior childlike Others, but are at times simply from a different, laboring class, then the civilizing mission is partially undermined. I am not arguing that my research participants' intermittent perceptions of Gilgiti servants as a different class, rather than a different race, guarantees anything like a reciprocal relationship with Gilgiti Others. However, by disrupting the justification of 'civilizing' imperial agendas, shifting perceptions of class and race can throw into question the concept of the civilizing self and hegemonic constructions of the 'primitive' racialized Other.

Before I proceed to the topic of child management, I return to the spatial organization of home, this time briefly to examine the distribution of household space between Western women employers and their Gilgiti servants and the consequences of those spatial divisions. During 'deep hanging out' (Geertz 2001) sessions in my research participants' homes, I noticed their spatial anxiety toward servants. Jane chose a distant working space once she knew where the ayah and cook had settled in to perform their chores. Amanda was fidgety when the guard, *chowkidar*, or "cleaning lady" was in the house. If she did not wake in time to be out before Safina arrived, then Amanda would try to avoid contact with her until a jeep was free for a trip somewhere else. Although Western women usually try to keep some distance between themselves and Gilgiti people, they have little choice but to have close physical contact with servants at home, especially with the ayahs who help raise their children. As it did in the colonial era (Bush 2004; Hall 2004; Procida 2002), this spatial proximity to employees disrupts Western women's efforts to achieve racial exclusivity. To reinforce a separation, my research participants often intensify their surveillance of 'incapable' servants. Despite the disciplinary authority that is experienced through unwavering supervision, this spatial closeness continues to disturb most Western employers. For instance, in a conversation that made me wonder why she ever decided to hire a servant, Amanda complains that "even when [Safina] is in the other room, this is too much for me. I want to be alone, but these [servants] don't understand. And then you get aggressive, and then they don't know what happened. But I don't want to be disturbed always when I'm sitting there doing something. 'Madame, madame' . . . She's *always* here, so I always have to *behave* so well." Even though she only comes a few times a week, Amanda experiences Safina as a constant, exaggerated presence in the house, a presence that frustrates her freedom to pursue indulgent whims and idiosyncrasies, such as walking around naked on hot days, wearing shorts, drinking alcohol before noon, and working all night. Her anxiety echoes that of Edith Cuthell, who in 1905 wrote, "The extraordinary lives we lead . . . unravel themselves like a ceaseless play before the unwavering eyes of our dependents" (as quoted in Procida 2002, 67). When servants are perceived as omnipresent nuisances who restrict Western women's lifestyles, they are experienced as dominant, controlling forces that need to be resisted. Amanda's experience is also reminiscent of white racism in Canada that confuses power flows by calling affirmative action hiring policies "reverse discrimination," as if the hiring of people from minority ethnic groups and their daily presence in Canadian workplaces is an oppressive pressure that unfairly victimizes white workers. Therefore, while domestics are valued by my research participants for their labor, they are also feared at times for the ostensible constraints their presences place over Western lives.

The living, as well as working, location of servants is another instance of the racialized division of household space. Most *chowkidars* live on or near the property, but are expected to enter the inner sanctum of the house only when they have indoor work to complete. As in Amanda's case, Esra's *chowkidar* lives in a hut near the front gate: "I was so lucky with my first *chowkidar*. He was so nice. He respected me and never tried to come in or look through the windows into the house. He was happy to live outside, in a small guard house in the garden.[23] I never felt uncomfortable with him there, not once. But if it was somebody else, I wouldn't be comfortable." Although Jane's sense of propriety makes nonfamilial men unwelcome in her home, her "manservant doesn't count because he just lives here [laughs]. He's OK. He works, but he also keeps me honorable when my husband's away. That's part of the reason why we have him. We built him a room onto our house. He does his own cooking, lives alone, he's more comfortable that way I think, and then he only has to come inside to work." The notion that Western and servant lives should remain spatially distant even within a situation of intimacy suffuses these comments.[24] I interpret this as an expression of my research participants' concern that intimate relationships with racialized Gilgitis in the home can be a further source of cultural contamination.

This racism is undercut by discourses of class that are exercised by nonemployer VSOs to disassociate themselves from what they perceive as imperial Western employers.[25] Most volunteers explain the division between themselves and long-term expatriates not only as a result of religious ideology, but also as a consequence of class practices. For example, Andy concludes that the social cleavage between Westerners in Gilgit "is a mixture of things. To be 100 percent honest, the groups are quite divided because VSOs come out here earning nothing and the other families do have big salaries. Many VSOs struggle with that . . . There is a class difference that revolves here around who has servants and who doesn't." Elena has little "to do with the other expats, really. It's always been a bit of a divide . . . It could be because of their church, or it could be that we're not earning big money and they sort of see us as different. It's odd. You wouldn't say we're all part of an expat community. There's a VSO community and then there's the monied expats." Joan is slightly bitter that

> all those development high ups don't have anything to do with me. They live in a *very* different community . . . Take the example of Jane's husband. He's on a million's wage higher than mine, and therefore he feels a social gap between them and me. It's class being played out. It's a *class* thing . . . He meets me in the office or I have meetings with him occasionally, but he wouldn't *dream* of socializing with me . . . I don't really take offense at it, but I interpret it as being a class thing, that these expats really know who's got the money, who's got the vehicle, who's got the house with everything, and they know that VSOs haven't.

Louise agrees that "there's definitely a Western divide. I mean, I can't invite Evelyn here, I can't offer her a beer or a fancy meal. She can come and have *subzi* [vegetables], she can come and sit on the floor and eat *hoi* [cooked spinach] . . . I've been to a couple of parties at her house, and there's fantastic food, a *chowkidar*, a cook. You know, we're *worlds* apart!" According to these volunteers, the Western class hierarchy is established on the basis of the quality of their food and homes, their salaries, the status and income of their spouses, and whether or not foreign households include servants.

Next to how monied one is, VSOs are most sensitive to and critical of the practice of hiring Gilgitis to do Western domestic labor. Susan still finds it difficult "to accept our landlord's *chowkidar* . . . Expats say 'Oh, he'll go to the bazaar for you, clean for you, cook for you.' NO THANK YOU! The idea of having a servant is alien. I know the British probably imposed it in the first place, but that's quite strange . . . There are things that I've said no about, like being waited on hand and foot." Andy's Western "friend has a gardener here, and I rib him the whole time about this. 'He's one of your servants. Send your servant IT around for me.' It's so colonial." Margaret is "not into the expat scene . . . These people haven't left home. There's a compound, they're living in it with lots of servants, mixing only with Western people. And that's not why I'm here. It's not good, not necessary to be so colonial." And when Louise sees "the houses of the expats and their big fancy cars, servants, swimming pools," she thinks

> "Oh god, look at this message" . . . It's incredibly embarrassing, but I think, honestly that's part of colonial history . . . They're reinforcing that image that *all* Westerners are colonials with loads of money to throw around . . . I think that's offensive, I really do. To be working for a development agency and living in such *fancy* accommodations. It doesn't seem to go together. I just wonder if British Council donors know that's how their money's being spent.

Volunteers accuse salaried Westerners of perpetuating colonial practices through the unequal, neocolonial labor relationship they have with Gilgitis as well as through conspicuous differences in levels of quality of life.

VSOs—who do their own cooking, cleaning, washing, shopping, and fetching—can act as disruptive forces in this once-colonized setting where white women have traditionally had servants to do this work. Their different class relation to Gilgitis as nonemployers who maintain far less decadent and 'cultivated' homes than salaried Westerners blurs some of the racial boundaries forged through the domestic middle-class representations and practices I have discussed. By doing domestic work themselves, even when they could afford to hire a Gilgiti person to do it, volunteers destabilize dominant racial relations together with the imperial legitimacy that depends on them. Antiracist impulses are also kindled when volunteers perceive indigenous class stratifications, which involve Westerners, as similar to what they know at home. Andy, for instance, claims that there are "very central ideas around here, much like at home, about who's the laborer, who's the educated person, and who's the guy who's made good. Like Iqbal, local people don't respect him even though he's a tour guide because he's still illiterate. So there's a big class system here too, and expats move into it at a tangent to the general strata, to the local class hierarchy." By acknowledging that Westerners and Gilgitis operate within similar class systems, sometimes simultaneously, Andy assumes a sense of social solidarity with indigenous society based on class relations, a sense of solidarity that may affect perceptions of racial difference and imperial legitimacy, because, as I have demonstrated, discourses of class, race, Orientalism, and imperialism so closely articulate in Gilgit.

Managing Western Families

Western women's efforts to govern their servants, like their representations of Gilgiti domestics, may be marked by ambivalent spatiodiscursive processes and effects. However, when it comes to managing the lives of their family members in Gilgit, Western women's disciplinary efforts are a far more straightforward matter of strictly upholding 'civilized' standards within Western homes. Even though this group of research participants strive to provide a carefully cultivated living environment for themselves and their families, family members require individual nurturing and "moral rearmament" (Bush 2004, 90) to maintain that environment. Margaret provides a good example of how she negotiates the 'cultural decay' that affects her relationship with her spouse: "Normally partners touch each other or hug when they pass. But because that's not allowed in *this culture* we're not touching in public. So I find we're not doing it at all, even in private, and you begin to feel like strangers, rather than a loving unit. That worries us . . . We have to make a point of at least kissing each other before we go to sleep every night to stop this problem." While spousal relations may be at risk in this cultural setting, Margaret demonstrates that Western women, together with their spouses, often manage that threat themselves by drawing on coping strategies that ward off the danger and preserve a comforting sense of self.

But immature children are not as adept at self-regulation and the "culture of monitoring" (Paisley 2004, 244). Despite the cultural benefits in a middle-class domestic atmosphere, children, according to Janet, must rely on the homemaker's discipline to keep them physically fit and culturally energized in Gilgit: "I can't even imagine how difficult it is for mothers, being responsible for young people here, because it's hard work keeping them fed, clean, getting school work done, keeping them healthy, energetic, and mentally active. That's a responsibility I would *not* want to take on. It would instantly increase my anxiety level." While spouses are often a concern, Western women's familial worry and disciplinary attention focuses primarily on children, developing their vigor, and thus sustaining the vitality of the Western home environment.

Considering these preliminary insights, I see the Western family in Gilgit, like the Western home, as a unit of civilization preserved through my research participants' efforts to avoid cultural contamination and, relatedly, to maintain domestic boundaries. Earlier in this chapter I discussed how the threat of cultural degeneration is both conceived and defused by Western women as they provide safe, healthy, and happy homes for their families. I expand that discussion here to show how other domestic practices pertaining to the governance of children establish racial and cultural exclusivity as a component of imperial power relations in the home. As a site where class and racial sensibilities are played out, Western families under supervision constitute yet another link between home and acts of imperialism.

The figure responsible for preserving Western families in the colonial contact zone has been the bourgeois mother, who is the custodian of her children's morality, health, and cultural fitness (Bush 2004; Chaudhuri 1988; George 1994; Hall 2004; Paisley 2004; Stoler 1995; Strobel 1991). Her strict governing of the family is essential in this setting, because "the family is where a child's sense of personhood, citizenship, and sexuality [can] be subverted, perverted, or well formed" (Stoler 1995, 152). The bourgeois mother's prime directive has been to raise a pure, vigorous 'race' to sustain imperial relations (George 1994; Paisley 2004). This maternal legacy is invoked unevenly and to varying degrees by Western women in contemporary Gilgit as they govern their children's lives in four main ways.

First, Evelyn, René, Abbie, Esra, and Jane are concerned about the consequences of their children's small social network in Gilgit. Evelyn, in particular, is worried that "even though the kids are happy with it, I don't think they have much of a social circle . . . They play with René's kids, Abbie's kids, Jane's kids, and a few other expatriate families, even in Islamabad. They know the kids next door. And that's it. It's fewer friends than they normally would have in Canada. Yeah, the lack of social variety may be a problem over the long haul." While Evelyn admits that this narrow social circle could be expanded to lessen her worry and possibly benefit her children, she is reluctant to expend the energy supposedly required to secure a new set of Gilgiti friends for the kids. Her diffidence is not a result of her desire to insulate the children from indigenous influence. However, by taking the path of least resistance, which is often paved with dominant practices and ways of thinking, this is precisely what happens.

René explicitly wants to protect her children from a 'restrictive' Gilgiti culture, visible in local youths and indigenous educational institutions: "In this sort of small, isolated culture, I'm really trying to provide my children with a very good education at our school, opportunities, and strong relationships with other foreign kids. I take that *very* seriously, it's a big responsibility." Jane is wary about her son interacting with 'destructive' Gilgiti children who have nasty habits and mannerisms: "It's a problem when [my son] plays with local kids. Children here don't play in the way our children learn to play. That's part of our culture of raising children. They'll play with stones and sticks that can hurt. They're not taught to be *kind*, or reprimanded for being *unkind*. So his feelings get hurt, he's vulnerable. And that doesn't happen when he plays with expatriate children." Jane is also hesitant to leave her son under the sole care of an ayah, whose disciplinary standards, like Gilgiti children's rules of play, may not be culturally appropriate for a Western child: "Although I wanted to work, I didn't want to leave my child with [the ayah] all day. He was cruising around, getting into trouble, needing some discipline and structure, and I felt that he wouldn't get the discipline and structure I wanted him to grow up with unless I was there to supervise him." Some 'discipline' is provided by hiring an English-speaking ayah who has worked part-time for foreign families in the past, because, as an Ismaili, she does not observe purdah.[26] This domestic hiring strategy is not only a matter of employer convenience, but may also be a way to confound the religious and linguistic 'perversions' that supposedly infiltrate Western homes via Gilgiti servants, putting children at risk. By isolating their children from Gilgiti youths and particular adults in these ways, Western women maintain exclusive racial and cultural boundaries to raise children with a Western sense of personhood.[27]

Second, my research participants' efforts to raise children with an appropriate sense of self include finding ways to keep children culturally stimulated. They also monitor themselves in this regard. Christine, for instance, reinvigorates herself through familiar music: "Rose plays the recorder too, and she's brought hers as well. Every Friday afternoon we meet up and play duets. So I do lots of playing, I listen to tapes, listen to the radio if I can find a BBC program, bake my own bread. That helps keep me sane . . . As I said, I also do a lot of reading. You really can't afford to let yourself go here by getting bored." Jane prefers her "life to be organized by a calendar, and a full one. I like to know that on Monday evening this is happening, Wednesday evening someone is coming to dinner. That doesn't happen in this culture . . . I can't *flow* with the randomness of things here. So I feel at *odds* with the culture, grating with the culture. I'm trying to loosen up a bit, but still keep up our cultural tradition." Evelyn illustrates that Western women are equally concerned about their children's cultural stimulation: "Books are *essential* here, not so many for me, but *definitely* for [the children]. So I bring hundreds of dollars worth of books from home to keep them stimulated all year. I also buy little presents for their friends so that they can always go to a birthday party and take a nice gift. They're off to another party tomorrow." Expatriate children are trained in appropriate behaviors by mothers who continue Western middle-class traditions such as avid childhood reading and hosting children's birthday parties complete with wrapped birthday gifts, guest grab bags, and homemade cakes topped with icing, imported candles, and candy sprinkles. These birthday celebration practices, foreign to Gilgitis, allow Westerners to avoid 'being' in the everyday cultural setting of Gilgit in which they physically live by pretending they have not left the West. This strategy reveals a fear of, or at least a perplexing lack of curiosity about local culture, which frustrates cross-cultural communications and foreigners' personal development, as it sustains imperial boundaries.

Western women cope with these family responsibilities, in part, by supporting one another. Evelyn explains that "Abbie's such a support to us mothers. All of a sudden in your house there'll be a jar of peanut butter for the kids, 'I thought you might need it,' that kind of thing . . . René, similarly, has an interest in my kids, 'Can your kids come over today at 4:00 pm to play computer games?' and I think, 'That's just what they need. That's very nice of you.'" While this interest in other women's children certainly reflects kindness, it also serves to remind those children who they are and where they come from by encouraging them to enjoy imported Western food and middle-class forms of entertainment with other foreign kids. Birthday parties, combined with large selections of books, stimulating leisure activities, proper friendships, and comfort food, may keep children in good spirits, but they also (re)inculcate those children into white, middle-class ways of life, and thus truncate their exposure to life as it is lived by indigenous people in Gilgit.

Third, once Western children turn five years of age, their mothers worry about their schooling. Esra explains that "children become a big issue for expatriates here. What are we going to do about our children's education? That's a huge issue, you know, wanting the best for them. Sometimes that becomes a focus of all your anxieties. It expresses our conceptual dedication. And it often gathers other things behind it." To ensure that Western children's upbringing abroad includes a proper, "conceptual" education, this group of research participants have decided to educate their children in a racially and religiously segregated setting either in Gilgit or down-country in Muree, where Christian missionaries operate a school for English-speaking Western mission children. The decision made by the four sets of parents to start their own private school was fairly straightforward, according to Evelyn, considering the lack of viable indigenous options—including the reputable army school: "Despite my worry, I think the kids are fine at our school . . . .Number one child took those British tests . . . He wasn't taught to those tests, but he's going off the scale, so I'd say, in some ways, being educated here by us hasn't hurt him. He would have more friends if we had made different decisions about schooling, but at the time there were no other options." Besides the local curriculum content and medium of instruction, Jane identifies one of the main problems with established Gilgiti teaching as

> this rote learning thing. It's just so different from the inspiring way our children learn and are taught from the earliest age by their parents. It's not good for them sitting in lines and rhyming things off by heart. Local teachers got frustrated, saying our kids weren't very good at learning things because they're not into that rote learning stuff. And the kids got frustrated because when they said they didn't understand *why* they needed to do something, the teacher said it wasn't *needed* to understand. You don't discuss it, you just learned it. So that was a difficult situation before we sorted things out in our own school.

By funding, operating, and teaching an alternative curriculum according to Western teaching methodologies in a segregated school, Western mothers can monitor their children's academic progress to ensure they receive, in their mothers' opinions, a "good," "inspiring" education free of indigenous influence.

Another problem arises when the children reach high school age, because the Western school curriculum ends at grade eight. At this point, mothers have the option of home-schooling their children or sending them to the Muree Christian high school, even though being separated from their children is painful. According to Abbie,

> I don't like at all having my son in a different city than us. But I'm really impressed with the Muree Christian School. It's a good school, good curriculum, they're well cared for, it's Christian, I like what they teach and how they train and discipline the kids . . . .I suppose the only alternative

would be to bring him back up here and do some kind of home-schooling. I just think that's far worse, because he doesn't get the independence and social interaction with kids of his kind . . . Now he's just so much more interesting, because he's got a bigger, richer world.

Despite their separation anxiety,[28] my research participants will send their children to school in the Punjab if they feel the institution provides a suitable cultural setting that instills appropriate Christian values and independence in their children. In what they believe to be the children's best interest, some mothers are even willing to consider sending kids, especially their daughters, back to Europe or North America for high school. René feels

it's important for my daughter to feel like she's an independent adult and learn how to negotiate the United States. If she grows up her whole life in Pakistan, and then we just sort of send her home for university, I don't think that she will feel comfortable as an independent human being in the States. So I think she should move to the United States for her high school years, just so that she can feel like, learn to *be* an independent and capable adult in the US.

These three schooling options enhance children's cultural capabilities, even within Pakistan, by ensuring they are taught Western ideas, values, and behaviors. This standard can be maintained only if mothers protect their children from indigenous culture, including teaching methodologies and religious values, that may influence their way of life. Western educational practices in Gilgit, then, are shaped by my research participants' cultural/racial boundary anxieties, as well as an imperial logic. Civilizational and racial exclusivity enacted through these educational practices inform foreigners' imperial subjectivities as they defend Western culture in Gilgit.

Finally, Western mothers attempt to cultivate a civilized sense of self in their children by closely supervising their physical vigor and health care. Like many middle-class mothers in Europe, Australia, and North America, my research participants value organized exercise for their children, especially the girls, whose ability to swim, rollerblade, dance, and play basketball at the Western school distinguishes them, according to these mothers, from ostensibly less healthy, homebound Muslim girls. Evelyn expresses her concern for her children's fitness:

You know what *torture* it used to be when we were all so hot. And now [the pool] regime has changed that *totally*. The kids aren't *bored* anymore, they have another regular healthy activity, they can amuse themselves. I think it's wonderful, the level of fitness the kids have here. If they aren't swimming, they're rollerblading. If they're not rollerblading, they've got a daily basketball game or they're walking the dog up the mountain. There's nothing they do that doesn't require fitness, so I hope that makes them rugged adults. I'm pleased about that. I always like *fit* kids.

By contrasting herself with her children and her friends, Evelyn describes how physical fitness and cultural discipline are mutually informing processes: "Most people here, including the kids, are very disciplined, so they wouldn't think of putting on six or eight pounds. Look at Abbie, she's very thin, and René has a figure to die for. They exercise a lot. But I'm not that *disciplined*. I feel like it's evidence again of how this place really gets to me." Warding off cultural deterioration seems to be, in part, a matter of sticking to a disciplined fitness regime. A disciplined body enables a disciplined sense of self.[29] However, that strategy is not always enough. According to Abbie, "I'm always tired despite trying to be healthy. On every visit to the States, I get my hormone levels and thyroid checked. It's always fine. I think it's just living here. It's exhausting trying to keep on top of *everything*, and I don't see what I can do differently. I could just lay around *a lot*, but then I think I would be depressed. Good habits help, keeping fit, the house clean, cooking."

Judging by Western mothers' behaviors and remarks, something as important to maintaining cultural fitness as physical health cannot be left to indigenous health care providers to protect due to their ostensible incompetence and lack of technology. René outlines the health conundrum mothers face when they decide to bring their children to Gilgit:

> My son broke his arm a couple of months ago. The health issue is not a joke, but I guess we feel like we just didn't know what else to do. So we agreed to live here, even when there is no *good* medical help. Two of our closest foreign friends here are doctors, so they can give good advice, but they don't have the facilities to do what they need to do in the case of an emergency. In an emergency you are pretty much exposed. You don't have any good place to go that can help you . . . It takes several hours here to get an x-ray and then they aren't really clear . . . So we had to drive down to Islamabad to the AKF hospital because the helicopter flight was full.

Because "good" indigenous care is apparently unavailable, Elena, like René and the majority of my research participants, prefers to consult foreign doctors: "I see the [American doctor] at the eye hospital. VSO now has an arrangement with him after hospital hours. I was taken to the local hospital once, but it was awful . . . I remember this guy, *supposedly* the doctor, he's got on a black leather jacket, jeans, x-ray in one hand and cigarette in the other. And I thought, 'Who are you? You're a nobody who just walked in off the street.' Never again!" When Gilgiti doctors' self-presentation differs so dramatically from the Western professional standard, it confirms many Western women's suspicions, like Elena's, that these are 'imposters' just off the street, returning from the stationery store where they purchased a photocopied medical certificate for the office wall. What mother could trust her children's health to these 'doctors,' even if many of them have been trained abroad, without supervision from Western physicians who live in Gilgit? These Orientalist representations of inherently incompetent Gilgiti physicians are also implicitly racist; they are predicated on the unarticulated notion that white Western doctors are more medically capable than indigenous physicians due to their racial, cultural, and technological superiority.

These views also guide my research participants' childbirth practices in Gilgit. The primary decision made around this especially sensitive health issue is to leave town. René's children were all born before the family moved to Gilgit.[30] Evelyn delivered her last child in Islamabad at the Aga Khan–funded Shaifa hospital. But the rest of my research participants who were pregnant in Gilgit, plus two Western women who are not officially part of my study group owing to pregnancy due dates, went back to Europe and North America to deliver their babies. Jane's first child was born during an educational stint in the United States, but her second was delivered "in Britain. Of course, I went home to my family and a good medical system. That's the longest time I was home since 1983." Few occasions apparently warrant lengthy visits back home, but childbirth is certainly one of them. As Jane suggests, my research participants return home fairly early in their pregnancies to find a pre- and postpartum support network they feel they lack in Gilgit. That support network includes familial attention, as well as good prenatal and delivery care that Gilgiti doctors and midwives ostensibly cannot provide. Lines of racial and cultural exclusion are therefore drawn around pregnancy, as well as pediatric health care. The lack of technology and racial and Orientalist discourses preclude health care from Gilgiti doctors and midwives, who are considered 'unfit' health protectors. If Western mothers are to guarantee the civility of their offspring, as well their homes, then all forms of cultural decay, including Gilgiti health and educational institutions and domestic servants, must be monitored and, in some cases, eschewed by a Western cultural milieu.

Domestic Governance and Spatial (Re)Organization

By examining domestic sphere activities, I have argued that the cult of domesticity and related practices of governance in the home constitute the feminine, middle-class, racialized, and imperial subjectivities of Western women in Gilgit. Initially, femininity may seem to be at odds with my research participants' understanding of themselves as liberated travelers and workers who aim to 'develop' Gilgiti society. However, domestic subjectivities are formed through intersecting discourses of authority *and* femininity exercised within domestic space. I have emphasized that, rather than reflecting 'public' imperial relations, Western women's domestic practices, especially those pertaining to servants, in fact constitute those relations. As racial and cultural boundaries are forged in imperial Western homes, domestic spaces become salient sites of social exclusion, as well as of identity formation and cultural belonging.

The cult of domesticity as it is practiced by almost all Western women in Gilgit usually involves processes of boundary management and political subjection with imperial consequences. These repercussions are *less* widely effected through women's surveillance of their servants and children, because most Western women in Gilgit neither hire indigenous domestic labor nor have children. And this group of development volunteers are often disruptive forces through their class and imperial critiques of salaried expatriates. VSOs undercut discourses of imperialism, hopefully to blur racial boundaries and to disrupt notions of cultural superiority. Consequently, some of the social practices that structure relationships of difference within domestic space are as enabling as they are constraining. When VSOs enact even slight shifts in domestic social relationships, they pave the way for equally important shifts in imperial power relations outside the home.

The significance to my research participants' identities of making and managing civilized homes in Gilgit refocuses our attention on the spatial and social processes that articulate to constitute subjectivity. I have described here how the spatial ordering of home is vital to practices and institutions of power. Western women's ability to assign places (and names) within the private, as well as the public, domains of Gilgiti life signals an exercise of power. This power to locate themselves, their children, and their servants in particular ways within domestic space is integral to broader processes of cultural valuing and (self-)identification. The making of the domestic haven, then, is also the making of social control and personal empowerment. Considering these aspects of the spatiality of power and subjectivity, we can see that creating new social orders and meanings will require simultaneous radical changes in the current arrangements of public and private spaces, because the social construction of space and the values, discourses, and material practices that (re)produce social relations are mutually constitutive processes: shifts in one social realm can effect a shift in the other.

Conclusion: Ruptures and Recuperations?

Epilogue

A postcolonial politic involves more than the dismantling of customary colonial institutions; it also necessitates the search for sets of discursive practices that resist imperialism in contemporary settings. A feminist sociology of imperialism can trace the vestigial and mutually constituting discourses and sociospatial boundaries that structure imperial practice even after the political independence of colonized states, an important first step in imagining and implementing alternative practices and more just social realities. My project has been to extend our understanding of the ruptured and recuperated relationship between Western women and imperialism over time by exploring the multiple and contingent discourses of power that organize a diverse range of Western women's daily practices and subject positions in contemporary Gilgit, Pakistan.

Successive chapters of this book have traversed the contours of Western women's subjectivity formation as it is discursively informed, materially practiced, and historically situated in the transcultural setting of Gilgit. Exploring these transcultural constructions of subjectivity and the ambivalent discursive repertoires that organize them has rendered more explicit some of the complex ways in which my research participants' self-imaginings, representations, and social practices are implicated in various operations of power in this setting, with effects at both local and global levels. My analysis thus opens a discussion about how Western women are positioned within the contemporary transcultural field of power relations in South Asia. Through this exploration I specifically question the dominant tropes that represent Western women as either the guiding force of imperial power relations or a benign presence in (post)colonial space (see Callaway 1987; Sharpe 1993; Strobel 1991). Moreover, I challenge the notion that experiences of imperialism reside in the past by identifying the relationship between currently circulating discursive frameworks and those exercised in the nineteenth and early twentieth centuries.

Social constructions of Sameness and Otherness are salient to understanding these contemporary operations of power. They articulate with geographies of exclusion within this transcultural field as my research participants fashion 'pure' spaces of home, leisure, and reproduction, free of contaminating Others. In other words, Western women's desire for exclusionary protective boundaries between Self and Other is realized both socially and spatially. Relationships between space, subjectivity, and power are exposed as their fear of difference becomes associated with specific places, like the bazaar. Feelings and desires are not always discursively organized (Kruks 2001), as my research participants' occasional tender feelings for their male Ismaili workmates and sporadic desires to interact with and respect Muslim women show. However, the anxieties of those women who are afraid to be in close physical proximity to what they understand as sexually and morally threatening Others are important aspects of exclusionary sociospatial relations in Gilgit. Western women's inability to fully segregate themselves from this danger prompts them to construct secluded safe places and to monitor transcultural interactions within these Western spaces.

In marking out some of the parameters of transcultural power relations in Gilgit, I suggest that Western women actively negotiate those boundaries as they (re)constitute their subjectivities abroad. I have explored the form and content of some of those negotiations in detail, by focusing on discourses of racialized sexuality and the 'oppressed Third World Woman,' as well as on constructions of home and the cult of domesticity. In chapter two I argued that my research participants are ambivalently situated 'at risk' in Gilgit according to discourses of racialized sexuality, which constitute white women as sexually vulnerable subjects and Muslim men as the source of sexual danger. To deal with the Black Peril, many of them avoid interactions with indigenous men and scrutinize their behaviors for signs of sexual excess. Most Western women feel less threatened by Gilgiti men when they can exercise some control over social, sexualized, and spatial interactions. These negotiations legitimate exclusionary boundaries that keep Western women and Muslim men hierarchically separated. Risky subjects thus spatialize the sociocultural landscape to produce spaces of inclusion and exclusion. In addition to highlighting the spatial aspects of Western women's subjectivity formation in Gilgit, I emphasized the importance of age and marital status as strategies they use to manage and reduce sexual danger. The continuing influence of Christian missionaries in this postcolonial setting, an influence that marginalizes many foreigners as well as Muslim Gilgitis, was another important insight. Some volunteers, including lesbians, who do not experience a sense of cultural belonging within the larger expatriate group due to this Christian influence, often reject the church community in favor of more tolerant Gilgitis who freely offer their hospitality. This sympathy for indigenous people questions 'us' versus 'them' sociospatial divisions and, therefore, may have some fracturing effect on Orientalist and racist power relations. Discursive disruptions are also enacted as my research participants begin to trust Ismaili men and understand them as sexually safe. This shift away from viewing all Gilgiti men as dangerous tempers their experiences of sexual and spatial threat, abjection, and cultural alienation, and thus partially disrupts dominant discourses of power.

Chapter three explored how Western women access discourses of gender and class in their travel experiences to develop a sense of themselves as independent individuals. My research participants perpetuate the construct of the 'liberated Western Woman' not only through travel, but also in relation to the 'oppressed Muslim Woman' whom they imagine as devoid of freedom and independence. Consequently, Western women's subjectivities are constituted through additional discourses of race and Orientalism pertaining to Other women. My research participants' clothing choices and perspectives on the value of their development work are infused with these discourses and constructs, naturalizing the hierarchy between metropolitan and Gilgiti women. The authority foreign development workers gain through local cultural improvement projects combine with their feelings of racial, cultural, and moral superiority to help them negotiate experiences of gender oppression. Clothing choices, as processes of identification and differentiation, also allow Western women to express and manage their subjectivities in relation to Gilgiti women. *Shalwar kameez* are used by my research participants to facilitate the marking of difference between modern, independent selves and primitive, dependent Others. The discourses of race and Orientalism that enable these identity binaries, however, sometimes conflict with Western women's self-understandings as wives, mothers, and homemakers. Christian women's middle-class domesticity opens a route to a possible sense of solidarity with Gilgiti women who are ostensibly homebound. This reciprocity undermines discourses of the self that posit Muslim women as radically Other. It also disrupts the notion that indigenous women are oppressed homemakers, because the lives of 'free' Christian women are similarly domestic. Those of my research participants who enjoy wearing *shalwar kameez* and refuse to see it as a sign of Gilgiti women's oppression are equally disruptive of discourses of race and Orientalism that discourage such a sense of solidarity. VSOs, however, have the potential to be the most destabilizing influence on these power relations through their critiques of metropolitan development initiatives as contemporary practices of imperialism.

My analysis of the cult of domesticity showed that Western women have recourse to intersecting discourses of gender, class, race, and imperialism to constitute their homes as havens from an Other world and spaces of cultural belonging that incorporate reassuring Western values, ideas, and norms of behavior. Most women manage a familiar sense of self in domestic space by hosting parties, preserving racial/-cultural distance and class respectability through privacy, and maintaining clean, well-appointed, safe, and happy homes. Married women also supervise their children and servants to create and sustain sociospatial boundaries of class, culture, and race. This making of imperial subjectivities and spaces allows this group of research participants to recuperate feelings of gender emancipation and cultural power developed through experiences of travel and development work, but confused by discourses of domesticity. But not all Western women claim such an imperial identity through the making of 'home.' Volunteers, who live less materially abundant lives than married women and as nonemployers have a different class relation to indigenous people, develop a self-critique of Western imperial practices that blurs some of the racial and class boundaries forged in the domestic sphere. Antiracist impulses are also kindled by VSOs who see indigenous class stratifications as similar to what they know at home. This sense of social solidarity may also affect perceptions of racial difference. Relatedly, when Western women view their servants as a different class of people more than a different race, their civilizing mission becomes hazier as the racial justification for it is undermined.

Perhaps the principal insight to take away from my analyses in these three chapters is that Western women who make temporary lives in Gilgit play multiple roles, which have contradictory and often unintended effects in this postcolonial context; different women come for different reasons, achieve different ends, and respond in varied ways to Gilgiti people, sociocultural circumstances, and lingering imperial practices. In their development work, as in their private lives, my research participants are neither civilizing goddesses nor imperial devils. As I have argued throughout the book, nothing is gained by flattening out ambivalent postcolonial power relations, especially into a tidy binary of neocolonizers pitted against the neocolonized. Transcultural relations in Gilgit are rather more complicated, fragmented, and blurred than that. Many practices of domination persist through a lengthy, though nonlinear, history, but they are at times disrupted as Western women question social and spatial boundaries. Dominant discourses and their effects thus comprise a complex process of rupture and recuperation, marked by a tension between disruption and recovery into newly negotiated discursive forms (Foucault 1978; McNay 2000; Stoler 1995; Young 1995).

The ambivalent imaginings and practices foregrounded in my analysis and reviewed here serve to highlight this process of potential rupture—or at least destabilization—and recuperation, and to locate where, within the limited disruptions, steps could be taken toward forging alternative social realities. Underscoring these sites of destabilization is an important part of understanding how imperial processes can be changed, even when these particular women have not used their sociospatial transgressions and challenges to logocentric thinking to realize definitive discursive ruptures and new self-imaginings. I return to these questions of rupture and transgression shortly. Ambivalence also provides evidence for the mapping of the complex structures of dominant discourses. Finally, beyond these theoretical and praxis-oriented concerns, emphasizing ambivalence has important epistemological and ethical effects. For instance, my research participants' contradictory identity performances and explanations extend avenues of interpretation, and they open the analytical arena to a plurality of conflicting voices. Writing a comparatively polyphonic text that incorporates these conflicting perspectives has helped me negotiate some of the problems associated with textual authority by challenging, in some small way, the conventions of representation used in traditional ethnographic accounts without abandoning them altogether.[1]

Three important avenues of further theoretical inquiry emerge from this project. First, in order to understand how Western women are implicated in power relations in Gilgit, we need some knowledge of the impact of their presence on indigenous people. How are discourses of power perpetuated and resisted through Western women's lives in Gilgit experienced by indigenous people? How are Gilgiti self-understandings, behaviors, motivations, perceptions, and life chances affected by the presence of foreign women? Second, and relatedly, to avoid Gilgitis remaining voiceless Others in a postcolonial analysis, it is important to learn their self-understandings about, for example, how they perceive foreign development workers or how they feel about *shalwar kameez*. Third, Nupur Chaudhuri (1992) has argued that British memsahibs served as a conduit for the flow of culture from India to Britain during and after the end of the Raj, and that the infiltration of a subordinate culture into a dominant one was a major effect of colonial rule. By importing Indian culture into Britain, mem-sahibs played a significant part in the global process of cultural diffusion and exchange, what Mary Louise Pratt (1992) has called 'transculturation.' As a study of transcultural power relations in South Asia, my project could be complemented by an analysis of what happens when Western women development workers return home. How are the relationally and ambivalently formed subjectivities and cultures produced in the postcolonial zone of cross-cultural contact lived, managed, and exploited in metropolitan settings, and to what effect? And how do they influence the ways the institution of development is reproduced in the metropolis?

Besides provoking new theoretical questions pertaining to the effects of Western women's subjectivity formation in Gilgit, this study also informs several broader issues currently debated in globalization literatures. First, within the body of research that considers landscapes of identity and groups of moving people in a shifting world, Arjun Appadurai's (1990) concept of 'ethnoscape' and Ulf Hannerz's (1996) categories of transnational flow specify five internally heterogeneous groups of people who have become principal features of our globalized world, affecting flows of identity, culture, and power (see Papastergiadis 2000). They include the transnational business class, tourists, refugees, immigrants and guest workers, and artists who invigorate the cultural scene. My project demonstrates that Western development workers need to be considered as one more significant, if not large, group of transnationals who profoundly affect global politics as they enable the flow of ideas, values, identities, and expertise from West to East or, to use terms more common in these literatures, North to South. These ideas and identities often become established as universals at the expense of indigenous ideologies, traditions, and agendas (Escobar 1995; Ferguson 1994; Mitchell 1995; Sutcliffe 1999). As the number of international development workers traveling to Africa, South and South East Asia, Eastern Europe, and Latin America grows, the global circulation of culture intensifies, and landscapes of power, especially in postcolonial spaces, are frequently (re)inscribed.

Second, and relatedly, there has been a conspicuous focus in the literature on cultural flows between metropolitan centers around the world, forming what many authors have called 'globalizing cities' (see, for example, Albrow 1997; Hannerz 1993; King 1995; Yeoh et al. 2000). Scholars often restrict their attention to transnational cultural flows between urban centers, primarily because they concentrate on migrating groups of people and businesses that (re)locate there. But once we consider the importance of development workers, we can see that rural areas such as Gilgit, where many development agencies are stationed and development initiatives implemented, are being globalized as quickly and intensely as urban centers around the world.[2] Globalization researchers need to reorient their focus to consider how rural landscapes of power are inscribed at the same time as urban centers through larger global processes.

Third, this analysis challenges much nonfeminist literature that assumes operations of transnationalism and globalization are gender neutral.[3] It supports feminist scholarship that explores the systematically gendered nature of material practices, spaces, discourses, and consequences of globalization and international development (Katz 2001; Mies and Bennholdt-Thomsen 2000; Nagar et al. 2002; Yeoh et al. 2000). These studies raise questions as to why only certain aspects of globalization—those related to men's lives—have attracted the attention of mainstream scholars, and how our understanding of those mechanisms is diminished as a result (Nagar et al. 2002). In an effort to address the politics of knowledge production at work here, subsequent feminist analyses will need to consider further how to conceptualize, study, and act in relation to a more diverse range of global operations, as well as to the gendered aspects of those operations already under study. To my mind, solving this problem entails building more grounded, contextual understandings of heterogeneous global processes.

Fourth, I have argued that imperial regimes of power continue to accompany transnational flows of metropolitan people and culture, particularly through international development activity. Hence, we should not think of globalization or transnational encounters without considering how they may be imbricated in historically variable and contingent acts of Western imperialism. Many globalization activities occur in a historical context of conquest, and even though that context may have been modified over time through assorted anti-imperial influences, its imperial past needs to inform analyses of contemporary global operations.

In addition to raising questions about globalization issues, my analysis of Western women's subjectivity formation suggests that some of the theoretical perspectives that frame my project need to be reconsidered. Two particular conceptual threads, which I will discuss in turn, pose problems for analyzing the data: the notions of (1) tactical social actions and (2) resistance as a perpetual potential embedded in everyday practices. First, although Certeau's (1984) theory of practice does not directly develop Foucault's rather nebulous notion of resistance, I suggest in the introduction that it strengthens Foucault's work in that it foregrounds the everyday practices—*les arts de faire*—that are not organized into discursive systems and regimes of domination, and thus are available to subjects as resources from the margins that can make daily life bearable within contexts of power. Certeau argues that these concrete everyday practices, which are creatively employed by social actors to achieve their own specific ends, are calculated tactics of resistance that insinuate an indecipherable difference into the structures of social life, although these processes of social adaptation are transient, unstable, and unpredictable. He reads this 'excess' of everyday life as fleeting moments of social openness when destabilized relations introduce the potential for new modes of knowledge and living. Resistance is thus an ever-present potential in everyday practices due to the social realm's undecided character.

While Certeau's insights into the imaginative, undisciplined aspects of tactical social actions provides a useful frame for theorizing and researching agency and lived possibilities for social change from the margins, these were only occasionally enacted by some women in my study, as when they affiliate and sympathize with indigenous people, accommodating and adapting to difference to initiate nonreactive practices. My evidence suggests that when Western women draw on resources to cope with life in Gilgit, they usually exercise, to varying degrees and often without intention, certain well-established ways of thinking and acting to negotiate others. For example, some of them perform exaggerated versions of femininity to avoid particular Gilgiti people, or enact imperial privilege more subtly. Many others deploy ethnocentric, bourgeois, Orientalist, and masculine discourses as strategies to negotiate male domination and resist gender oppression.[4] Consequently, while some acts of 'making do' pertaining to the clothes they wear, their homes, and their relationships with Gilgiti people do disrupt dominant social relations, most others do not help alter structural constraints or create less oppressive networks of power.

It seems that in order for my research participants to express and sustain comforting identities, they need to establish a fairly coherent narrative of self, even though "such practices of representation always have to negotiate the divided field of enunciation: the subject who speaks and the subject represented in its narrative never quite coincide" (Barnett 1997, 140). In order to act autonomously and invest their uprooted lives abroad with continuity and flow, my research participants construct an intelligible sense of self through stories they tell themselves about themselves. I suggest that these narratives also serve as practical guides to behavior so that Western women usually do not behave in ways that unsettle their self-representations. Although some women learn to act and think about themselves differently in Gilgit, others' self-imposed behavioral and conceptual restrictions, as well as external hegemonic social norms, help sustain their subjectivities and limit avenues for social and self-transformation.

I emphasize, for instance, that even though my research participants occasionally are disruptive in their clothing choices and interactions with Gilgiti women and men, their efforts at change tend to be organized around their feelings of oppression. In other words, disruptions usually occur when Western women are self-assertive, because they understand themselves as the gendered 'oppressed' far more often than the ethnocentric, racist, or bourgeois 'oppressor' in this Muslim setting. As David Spurr (1993, 195) notes, "Western subjectivity is powerful enough to absorb a certain amount of interference," so when my research participants heed Other voices or perspectives at all, either dissenting or supporting, in the process of relational subjectivity formation they are often co-opted into Western narratives as enhancements. Gilgiti women's complaints about their "not always good" husbands, for example, enlarge Western women's self-understandings as liberated women who become downtrodden in Muslim cultural settings, rather than signaling a shared experience of gender oppression. Strong narrative identities prevent many of my research participants from examining their own interests in maintaining certain power arrangements. Western women often feel trapped within gendered structures of power in Gilgit. Therefore, they are not usually sensitive to the limited and contingent nature of dominant discourses of power they exercise to negotiate their gender oppression.

The loci of their few disruptions are also locally and historically contained sites. For instance, *shalwar kameez* has special social significance for many Muslim Pakistanis as a piece of clothing that survived colonial occupation and signals Muslim respectability and continued resistance to contemporary Western cultural imperialism, especially to norms of dress for 'immodest' women (see Cook 2001). Western women risk being seen by indigenous people as morally lax if they do not dress 'properly,' with negative effects for their development work. This imposed lack of choice regarding clothing reinforces discourses of 'oppressed Muslim women' and 'liberated Western women.' Consequently, even when a few of my research participants are not bothered by wearing this garment, there is little room for reconstituting Western selves within this particular locus of discursive disruption.

Second, considering that Western women's practices of 'making do' often reinscribe social relations instead of destabilizing them, I want to examine further the notion of resistance as a perpetual potential embedded in everyday practices. When is potential realized as resistance? How do we describe moments of social destabilization (and their effects) that ultimately resolve back into dominant social forms?

Many scholars would argue that whether the potential for resistance is realized or not depends on the *effect* that a given practice has on the nexus of power relations in which it is embedded. So, a practice is resistant if it alters established social structures to lessen their dominating effects. However, as most social practices have unintended consequences, James Scott (1986) contends that any definition of everyday resistance requires some reference to the *intentions* of social actors. Many theories of resistance imply some intention and capacity on the part of subjects, including those of Foucault (1978), Butler (1993, 2004), and Certeau (1984), particularly when they speak of subjects actively redefining their bodies, reconfiguring their desires and attachments, and creatively consuming arts and crafts that make life tolerable and livable. Everyday acts of resistance, then, are defensive forms of self-help intended to mitigate exploitation, appropriation, or dominant norms, although they may only marginally affect these forms of domination (Scott 1986). Foucault (1978), for example, sometimes frames resistance as a 'reverse' discourse that bounces back to undermine particular aims of normalization, and thereby suggests that resistant action is directed against an identifiable obstacle with the purpose of changing it or lessening its effects. The goal of everyday agonistic resistance, according to both Scott (1986) and Foucault (1978), is not to overthrow or transform a system of domination, but rather to survive within it for a while longer, to endure to fight another day.

While I sympathize theoretically with the notion that resistance to forces of domination involves the intentions of social actors to change some particular obstacle or advance their own claims, my empirical findings suggest that Western women in Gilgit, for the most part, do not refashion their subjectivities and social structures this way. In thinking about how to identify the potential for resistance and how to describe moments of social destabilization in this setting, I find Tim Cresswell's concept of transgression, which is commensurate with my theoretical emphasis on practice, more helpful than resistance in highlighting how subjectivities are reconfigured through boundary negotiations.[5]

Cresswell (1996, 9) defines transgression as a "form of politics that questions the normative world by underlining what values and norms are considered correct and appropriate." It involves an opposition to established norms that contests the hegemonies of everyday life by confounding 'common sense' understandings of the world through social and spatial boundary crossings. Acts of transgression and resistance are connected in a complex relationship. But while both contest topographies of power, transgression destabilizes social relations through mainly incidental boundary challenges. In contrast, intentional, agonistic practices of resistance are characteristically achieved by incrementally altering networks of power. Transgression provides some potential for resistance and social change by breaching what is considered 'normal' and questioning, in practice, if not analytically, taken-for-granted aspects of life. For this reason transgressive acts appear to be 'against nature.' In Cresswell's (1996, 26) words, "derivations from the dominant ideological norms confuse and disorient, and therefore reveal the historical and mutable nature of what is considered 'the way things are.'" Boundary challenges, however, do not necessarily lead to intentionally resistant acts that forge new modes of living and knowing. But they can engender new ways of thinking and being if they are not resisted in turn.

For example, as I show in chapter two, some Western women in Gilgit often incidentally transgress sociospatial boundaries of race and sexuality that involve indigenous Muslim men in the public domain at the local scale.[6] While they are vigilant to maintain inviolate private spaces and interactions, outside this sphere my research participants find themselves working with indigenous men, teaching them, buying clothes from them, and sitting beside them in Suzukis, all, admittedly, to little ill effect. Moreover, men's hospitality convinces some Western women to drink tea with them in public cafés and to accept offers of assistance and dinner. And they, as well as the other women witnessing these boundary crossings, are often surprised when their fears of vulnerability are not realized. Hence, some of my research participants begin to understand particular Gilgiti men and local spaces more positively, which begins to throw into question for them hegemonic discourses of sexual vulnerability. Western women also transgress racial and imperial boundaries when they perceive some social connectedness to the lives of indigenous women, including mutual experiences of sexual harassment, work, family life, and gendered respectability.

This practical questioning of the taken-for-granted binary liberated Self/oppressed Other, like that of the generalized character of women's fear, contests hegemonies in such a way that Western women realize only the *outcome* of these transgressive actions and representations, but have not intended to resist an identifiable oppressive obstacle. Moreover, these boundary challenges, when realized, are sometimes 'corrected' as women enact reactionary ways of 'making do' that reinscribe established discursive systems. At other times, however, boundary challenges are not corrected, and then boundaries shift somewhat. Acts of discursive recuperation in the moment of social destabilization exemplify that we cannot have a priori knowledge of the meanings of transgression or how those meanings may be combined with others to change or sustain the conditions of social life. Transgressive acts are contingent in that they may not all be equally subversive or even potentially so. Transgressors' responses to their own transgressive acts need to be the focus of interest: will social actors seize the opportunity presented, use the potentially resistant opening in social life from which new knowledge can be formed? The political significance of transgression, like resistance, depends on what kind of knowledge emerges from the site of social disruption.

I contend, then, that Western women often enact a politics of transgression rather than resistance, because their boundary challenging behaviors and representations are neither intentionally activated nor agonistic. Hence, disruptive effects often attach only to the effects of those particular behaviors, on what the actor and other witnesses notice after the boundary challenge (Cresswell 1996, 23). In other words, my research participants often unintentionally cross lines meant not to be crossed, and these transgressive crossings are arbitrated by those who react to them, including the transgressor, in a social accounting. Western women's boundary crossings are always in reaction to some technology of power, but they are not always ways of creating new social and spatial configurations from the margins, as Certeau envisions everyday practices of tactical resistance to discursive regimes. So while the effectiveness of transgression lies in its ability to explore multiple topographies of power in the form of common sense assumptions about 'the way things are,' its limitations lie there as well. The constant destabilizing of sociospatial boundaries may not lead to gradual or sudden social transformations, especially when my research participants' ambivalence often purifies or explains away their experiences of transgression, particularly in regard to Muslim women (see chapter three). Transgression thus involves the temporary disruption, rather than the radical rupture, of discursive imperatives. Transgressive boundary challenges need to become more regularized or institutionalized in order to effect further, lasting social changes. Western women need to build on incidental disruptions to alter structural constraints and realize a less passive agency, one that actively rejects imposed symbolic categories and boundaries, exercises different discourses, and lives more comfortably with unknowing.

Agency Speculations

My final aim here is to situate the Western women in my study more precisely than I have done so far in the complex web of power relations enacted in the cross-cultural situation of Gilgit. Following a scheme of subject 'profiling,' I want to consolidate the main discursive strategies perpetuated and destabilized by my research participants into what I foreshadowed in the introduction as a theory of agency. However, I hesitate to designate the product of these last few pages as a 'theory,' due to the universal and monolithic connotations associated with that concept. In addition to these epistemological concerns, I cannot unequivocally locate Western women in a field of local and global power relations in Gilgit, mainly because of the ambivalence of processes of subjectivity formation. However, there are some striking behavioral and representational commonalities among many of these women, rooted in a centuries-old discursive history, that are worth summarizing into what can be more modestly called speculations on 'her' agency. To prevent commonalities from overshadowing differences within and among women in this profile, I want to stress that not one of my research participants conforms to all of these speculations, and that there are important outliers.

The exploitative aspects of the uneven history of Western colonialism and imperialism in South Asia may not be 'her fault,' but she inherits this history primarily as a beneficiary in the postcolonial setting of Gilgit. It shapes the present in which her life is discursively and materially lived in Gilgit, situating her in a range of sociospatial relationships that include privilege, distance, segregation, and local, fragile, and situationally specific forms of quasi-integration. She lives, negotiates, appropriates, and sometimes destabilizes, more unintentionally than not, an entire array of discursive repertoires that coexist in the present in uneven and complex ways.

In creating a life abroad, she ambivalently fashions her own subjectivity using the discursive materials at her disposal. Despite 'the way things are,' she adopts the adventurous life of travel and development work as a means to shape an independent female identity and often to refuse the Western vision of the feminine, even as she believes that such a vision is grounded in gender equality. Her work allows her to negotiate local social restrictions for women, although the masculinist development hierarchy limits her opportunities by relegating professional women to subordinate positions. She searches for every opportunity to lead a nonfeminine social life in Gilgit, yet her life is primarily domestic after the working day is over and spatially confined, in part, by virtue of her own choices and preferences.

'Against nature,' she is also periodically able to accommodate and adapt to cross-cultural differences through acts of transgression. Occasionally she breaches borders of culture, race, and sexuality to recognize some limited parity with Gilgiti men and women. At these times she acknowledges a common three dimensionality, relating to indigenous people more as 'subject to subject' than 'subject to object.' A dialogue of quasi-reciprocity ensues when she admires Gilgitis, learns a local language, and recognizes similar class systems across cultures. Self-mockery and feelings of delicacy toward Gilgiti people, together with moments of sympathy for them, soften social boundaries to accommodate some degree of difference. Similar results are achieved when she enlarges her imagination, tolerates ambiguities, views Muslim men's staring as a curiosity akin to her own, criticizes foreigners for perpetuating imperial practices through international development and servant and household management, and does not lose sight of the interesting details of everyday Gilgiti life.

But history exerts a strong influence on her discursive performances in this part of the world. Drawing on racial and civilizational discourses, she often hesitates to assign full humanity to Gilgiti people, unlike the way she assigns full humanity and complexity to herself. When she employs indigenous servants, she tends to treat them as though they were children. And she generally argues that, due to 'barbaric' social customs that degrade women and Gilgiti men's 'primitive' sexual desires, 'civilizing' Western interventions, enacted through development initiatives and personal example, are required to compensate for the cultural/racial inferiority of 'natives.' When she adheres to the axiom that the 'West is best,' she perpetuates discourses of indigenous incompetence and Western cultural superiority, to reinforce imperial authority and conquest in the guise of international development. She achieves racial distancing—be it in the bazaar, at parties and the pool, or in her home—when she deliberately avoids interacting with indigenous people and learning from them, and when she is incapable of speaking to them in their own language. If she has children, these incapacities and avoidances extend to them, as they are sent back to Europe or North America for an education or enrolled in white-only expatriate schools in Pakistan.

She frequently exploits racial inequality to realize more freedom as a feminine subject-cum-honorary man. That inequality is forged partly through racial divisions based on the making of white bourgeois morality, sensibilities, and sexuality, all of which require a carefully prescribed, 'civilized' home and school milieu for her and her family. She also capitalizes on socioeconomic and racial inequalities when she employs indigenous servants to do hard domestic labor for low wages. Constructions of race constitute Gilgiti people as biologically suited to this particular kind of menial work, translating discourses of racial superiority into class terms. Economic disparities are thus maintained through discourses of racial distinction merged into class distinction.

She tends to draw on her femininity to get what she needs, when she can. She also performs a feminine subjectivity, perpetuating dominant discourses of gender, through her economic insecurity as a volunteer, or her financial dependence on her spouse. In the arena of development work, she is occupationally segregated in the gendered realms of education and health care, nurturing Gilgiti adults and children, as well as her family. She fulfills her philanthropic, familial, and religious duties through this multidimensional bourgeois benevolence. She frequently understands her responsibilities to include uplifting the 'oppressed' Muslim women of Gilgit through her work and superior example. When she has a family to nurture, her exemplary domestic management can shape current manifestations of imperialism through the governance of her children and her servants.

Replicating a bourgeois sense of self in Gilgit helps her achieve a familiar identity abroad, as well as some experience of agency. If she is a mother, wife, and homemaker, then class privilege is often manifest in her large and well-managed house with its Martha Stewart décor and upscale amenities. In this setting, servants help organize her gracious hostessing responsibilities, including house parties, anniversary dinners, and Saturday evening meals after the church service. In keeping with bourgeois discourses of domesticity, she associates herself with the private sphere as a moral guardian and repository of moral values for her family. As such, her self-sacrificing philanthropy is linked to her domesticity. As a volunteer or longer-term resident, bourgeois discourses also inform her choice of friends, social activities, spatial movements, and type of work in this social setting. Through her social reforming activities and moral guidance, she often hopes to provide a civilizing influence on Gilgit's inhabitants. She is prone to juxtapose to this 'faceless mass' her own unique bourgeois individuality, legitimating her mission as a traveler to know and represent the Other.

When she fails to understand that Muslim Gilgiti people operate within a vital sociocultural system and lead complex lives that operate according to their own logic, she perpetuates discourses of imperialism that position the Euro-American white Self as central and the not-white, non-Western Other as a one-dimensional observable object without subjectivity. As objects, they can be infantilized and sexualized. She also enacts imperial power when she accepts the authority and superiority of everything Western, especially in those moments of threat when she feels the absence of Western civilization most. By teaching Western morals and culture, she hopes to transform Gilgitis into something more than they have been. She enlists the trope of rescue when she, as an agent of modernization and development, aims to save indigenous women from cultural disorder. But she may also be transformed through her work; she can use her experience in developing Others to gain knowledge and occupational specialization, which may translate into professional advancement, work autonomy, and pay increases back home.

Although the results of this analysis, as a text, are open to rewritings and rereadings in discussion and debate, analyzing the connections between Western women's daily lives in Gilgit and various discursive regimes may have made visible some of the processes by which technologies of power are secured, reproduced, and destabilized during contemporary transnational encounters. In this case, tracing acts of imperialism into the present and prefiguring discursive landscapes are political acts in themselves. Ultimately, the process of altering present and future meanings of any one aspect of subjectivity is inextricably connected to that of altering the meaning of other mutually constituting discourses and sociospatial boundaries. That process, in turn, is linked to the effort to transform discursive terms and the distribution of socioeconomic power, and to alter the global circulation of power. While that is certainly a long-term collective project, including the actions and representations of people from a range of locations within various fields of power, I have attempted to point out where and how individual subjects, in their daily lives both at home and abroad, could take the opportunity—once they have done the very difficult and challenging work of reflecting on their own privilege and inculcation in a diverse set of discourses in operation there—to redefine and reorganize local and global power relations as they reconfigure their subjectivities and daily practices, as well as established sociospatial boundaries.

Notes

Introduction   Points of Arrival
and Departure

1. Pseudonyms chosen by individual research participants and friends are used throughout the chapters.
2. Ismailis comprise the third largest Islamic group in Pakistan. In the eighth century this group parted from the Shia mainstream to follow a different son of the sixth Imam. The present Imam, Prince Karim Aga Khan, is forty-ninth in that hereditary line of spiritual leaders. In the eleventh century Nasir Khusro, the famous Persian poet and philosopher, brought Ismaili teachings to the Hindukush region (Dani 1991). Most non-Ismailis in the Northern Areas associate the Aga Khan with the Aga Khan Rural Support Program and other development agencies funded by the Aga Khan Foundation.
3. The AKDN has expanded its activity in the area by funding the Professional Development Center at the Aga Khan University in Gilgit, Aga Khan Health Services (AKHS), Aga Khan Educational Services (AKES), Aga Khan Trust for Culture, and Aga Khan Planning and Building Services (AKPBS). AKPBS currently has two high profile projects running out of Gilgit. The Building and Construction Improvement Program (BACIP) began in 1998 to improve local built environments, and the Water and Sanitation Extension Project (WASEP), which acts as a complement to BACIP, began to provide water and sanitation services in the Northern Areas in 1999.
4. For example, the Gilgit Eye Hospital opened in the late 1980s, soon after AKRSP arrived. This clinic began as part of British Vision International, but now the cataract surgeries and intraocular lens implants are performed by American doctors and funded by Earth Mission, an American missionary NGO. The World Conservation Union arrived in Gilgit in the early 1990s, as did the United Nations–funded Ceena Health and Welfare Services and the Al-Amin Foundation, both of which are concerned primarily with AIDS prevention. The British Council–sponsored Northern Areas Educational Project (NAEP) was initiated in 1998. With funding from the World Bank and the Department for International Development (DFID), the British Council aims to promote the United Kingdom as a positive force in Pakistan by encouraging educational reform in the Northern Areas, especially girls' access to education. Finally, the Commission on Science and Technology for Sustainable Development in the South (COMSATS) began to provide an internet service in Gilgit in the fall of 1999, in large part to facilitate development agency activity and communications.

5. I use the term 'Western women' to describe my research participants as a social group. I acknowledge that group designations in general (e.g., Gilgitis, foreigners, expatriates) are problematic in that they can be ambiguous, homogenizing, exclusionary, or overinclusive. With these difficulties in mind, I employ the term in keeping with what is most commonly used by my research participants to describe themselves, while being specific about their 'home' locations. Moreover, my analysis of processes of subjectivity formation in Gilgit outlines the complex differences among and within individual women of this group, which goes some way in undoing the term's homogenizing tendencies.

6. Chris Weedon (1987, 32), arguing from a poststructural feminist perspective, refers to subjectivity as the "unconscious and conscious thoughts, emotions, and actions of the individual, the sense of herself, and her ways of understanding her relation to the world."

7. The term 'postcolonial' has multiple meanings. Here I refer to the formal political status of a sovereign nation-state that is no longer a colony. However, this usage can be misleading. Restructured political arrangements may have undone overt colonization, but many citizens in politically independent states are still affected by various types of Western domination that were first created under colonialism. This situation demonstrates the limits of the term 'postcolonial' as it is used in this sense, particularly the unfounded optimism implied in the prefix 'post.'

8. Other works that have explored Western women's varied roles in Empire include Bulbeck (1991); Clancy-Smith and Gouda (1998); Inglis (1975); Levine (2004b); Midgley (1998); Pierson and Chaudhuri (1998); Procida (2002); and Rafael (1995). For overviews of this body of literature, see Buzard (1993b); Formes (1995); and Haggis (1990).

9. As opposed to a nation-state's political status, here 'postcolonial' refers to an academic field of study and a theoretical and political position concerned with analyzing how imperialism and colonialism have been practiced (mainly by the West) and what has been done to resist these processes through anticolonial social movements in Africa, Asia, Latin America, and Europe.

10. The term 'subaltern' was originally used to describe a British military officer who had a rank lower than captain. However, Italian theorist Antonio Gramsci used 'subaltern' to refer to subordinated groups marginal to established structures of political representation. This meaning has been taken up (and made problematic) by postcolonial theorists such as Gayatri Spivak (1988) to describe (once) colonized populations that have been denied most forms of political agency by colonizers.

11. See Jayawardena (1995) and Procida (2002) for more complex descriptions of the competing interests and multiple roles of Western women living in colonial India.

12. Consequently, my use of the term 'community' invokes neither the normative ideal designating how social relations ought to be organized (see Tönnies 1957) nor a metaphysical logic of identity that conceptualizes

community as a unified whole (Young I. M. 1990). Rather, I use 'community' in the sense of a small group or social network whose members share some common characteristic—such as being women and foreigners—in addition to a common location.

13. Only one woman refused to participate, and I missed four others who were either out of the country for the summer or remained unknown to me until the last minute.

14. See, for example, Butler (1990); Diamond and Quinby (1988); Foucault (1977a, 1978); Fraser and Nicholson (1990); Frye (1983); Fuss (1989); Haraway (1985); Hekman (1996); hooks (1984); Irigaray (1985); Kristeva (1981); Lorde (1981); Lyotard (1984); Mohanty (1984); Nicholson (1990); Pile (1993); Rich (1980); Spivak (1985, 1988).

15. By arguing that "a social position involves the specification of a definite 'identity' within a network of social relations" (1984, 83), Giddens cannot theorize an unstable subject with multiple and conflicting identities who, when situated within a specific social circumstance, practices its identity in ambivalent and contradictory ways. True, his reflexive 'I,' originally formed in the mirror phase of personality development, may shift its "positioning in the time-space paths of day-to-day life" (1984, 85), especially along the continuum of the life cycle, and change its *category* of social identity. However, this 'I' maintains a 'definite' and stable identity in each phase of the cycle.

16. Giddens also describes social actors' knowledge as neutral; accurate understandings and explanations of social life are formed autonomously from the fields of power that mediate social action. By positing the construction of knowledge outside the constraints of power and the context of social relations, Giddens again unravels his recursiveness. Agents reproduce structure when they draw on certain social conventions, but their knowledge of social relations remains separate from the power that constitutes those relations. In this way structuration theory reproduces the notion of neutral knowledge as critiqued by Foucault (1977b, 1980, 1984).

17. Bourdieu (1990, 54) concedes that "the division of labor between the sexes, household objects, modes of consumption and parent-child relations" are important primary socialization experiences; but he remains committed to the Marxist vision of class as the primary force of socialization and social inequality. Indeed, by theorizing class as the "conditions of existence" (1990, 54, 58, 59), as opposed to a social production location, he can claim that "class habitus" "manifests" itself in experiences of gender, age, and ownership (1990, 54). By suggesting that class habitus constitutes an identity that is active across fields and manifests all other aspects of identity, Bourdieu posits a mobile, yet unidimensional class-based subjectivity. He (2001) later attempts to include a gender dimension in his notion of subjectivity by examining masculine domination, but, as he never really explains what he means by masculine domination and how it constitutes gendered subjects, he does not successfully disrupt his one-dimensional social actor.

18. Despite his shift away from some Enlightenment ideals, Foucault did not jettison appeals to Enlightenment values such as reason, autonomy, and dignity. He (1984, 43) argues that we should not "be 'for' or 'against' the Enlightenment . . . one has to refuse everything that might present itself in the form of a simplistic and authoritarian alternative: you either accept the Enlightenment and remain within the tradition of its rationalism . . .; or else you criticize the Enlightenment and then try to escape from its principles of rationality . . . And we do not break free of this blackmail by introducing 'dialectical' nuances while seeking to determine what good and bad elements there may have been in the Enlightenment."

19. If a particular discourse is not circulating in a social setting, it will have no effect.

20. Recurring metaphors, analogies, binary oppositions, apologies, and efforts to differentiate social groups, set rules and standards, and determine what is normal or pathological are, for example, important discursive regularities.

21. See also Heron (1999) and Parfitt (1998). Relevant studies by Goudge (2003) and Kealey (1990) include both women and men who have worked abroad in the field of development.

22. I use the term 'transnational conceptual frameworks' to refer to "the reordering of the world through forms of knowledge reworked from their entanglements in longstanding coercive power relations . . . activities by which new subaltern histories, new identities, new geographies, new conceptualizations of the world—transnational rather than Western—are fashioned and performed, and seeks through them to redress current imbalances of power and resources in the pursuit of more just and equitable societies" (Young 2001, 66).

Chapter One   Bazaar Situations

1. *Subzi* is the Urdu word for vegetables.

2. Of the 14 languages spoken in the Northern Areas and the bazaars of Gilgit, at least half are indigenous. The remainder is spoken by Central Asian traders (Dani 1991).

3. Purdah is the Arabic word describing the spatial, sexual, and social segregation of women in the home that is practiced by some Muslims. It is a system of rules and norms of behavior for women, which aims to prevent or reduce interaction between men and women who are not kin-related.

4. *Het* means 'neighborhood' in Shina. The plural form of the word is *heti*.

5. Gradually, over 15 years, Uma has expanded David's sphere of movement in her house. If we are the only nonfamilial guests, she serves dinner in the inner courtyard near the kitchen, and Uma will eat with us, although she sits at the edge of the room.

6. *Shalwar kameez* is the traditional Punjabi outfit consisting of a long tunic (*kameez*) over baggy pants (*shalwar*) that is worn by most women

throughout Pakistan. A *dupatta*—a long scarf that is worn over a woman's shoulders and chest or over her head as a form of veiling—is an integral part of a woman's *shalwar kameez* ensemble. *Shalwar kameez* is understood as requisite clothing for women (but not men) in Pakistan, because it enacts and respects Islamic codes of modesty, morality, and respectability.

7. Most formal health care in the Northern Areas is funded by development agencies, not by the Pakistani government. The Aga Khan Health Service, for instance, funds basic medical training for interested and educated Ismaili women. These Lady Health Visitors provide front-line first aid, diagnosis, and birth control information and treatment in villages, where doctors are seldom located. Lady Health Visitors were initially a late-nineteenth-century phenomenon in Britain and France. They were used by local authorities, in conjunction with women's volunteer organizations, to improve the quality of the 'imperial race' by monitoring the health and domestic life of the urban working-class population (George 1994).

8. A burqa is a large piece of fabric—often shaped like a shuttle-cock and made of brightly colored cloth—that some Muslim women wear in public to cover themselves from head to foot. Loosely woven areas in the fabric around the eyes enable women to see. The burqa is the most 'complete' form of veiling, but it is seldom worn by Gilgit women.

9. Drew is undoubtedly drawing on, indeed citing, particularly sympathetic insights made by previous travelers to the area, most notably as recounted in texts written by Dr. Gottlieb Wilhelm Leitner (1877, 1889) who in 1866 was the first European to record his visit to Gilgit. And Drew's book was certainly read by John Biddulph, the first political agent in Gilgit, the explorer Francis Younghusband (1892), E. Knight, a traveler and correspondent, and Algernon Durand, the political agent stationed in Gilgit in the late 1880s. These later imperial travelers, eager to acclimatize themselves to the borderlands of the British Raj before setting off on their own adventures there, read, and often cited in their own writings, the authoritative counsel, recommendations, and cautions issued by their predecessors. Texts written by Bellew (1875); Conway (1894); Robertson (1896); and Vigne (1844) have also been influential in representing this space of Empire to European outsiders.

10. Texts written by Cockerill (1922); Curzon (1927); de Filippi (1915); Earl (1930); Etherton (1911); Mason (1931); Morrison (1904); Schomberg (1935); Shipton (1938); and Workman (1901) cite many of the same imperial experiences, observations, and interpretations of the Northern frontier, and thus also contribute to the legacy and legitimacy of nineteenth-century representations.

11. Sarah Mills (1994, 41) argues that European women travelers in the colonial era often took great care to produce "knowledge of a very particular and safe feminine type," including descriptions of flowers, butterflies, folk tales, and local customs. Although they practiced this type of femininity to create an apparently neutral, safe, and unthreatening feminine

persona for themselves, their writings are actually imbricated in colonialism by representing the colonized country as a storehouse of collectable 'specimens' and traditions that civilized metropolitan modes of thought can order and thus render comprehensible.

12. This gender difference in Western attitudes about indigenous people reflects the dominant colonial representation of the colonized territory as female and the colonizers as male. See chapter two.

13. James Mills, in his *History of British India* (1818), provides a succinct overview of the imperial argument that women's social status marks a society's level of civilization and progress: "Nothing can exceed the habitual contempt which the natives entertain for their women . . . They are held, accordingly, in extreme degradation. Among rude people, the women are generally degraded; among civilized people they are exalted. As societies advance, the condition of the weaker sex is gradually improved, till they associate on equal terms with men, and occupy the place of voluntary and useful coadjutors" (as quoted in Hall 2004, 50–51).

14. See, for example, Band (1955); Michaud and Michaud (1975); Schor and Schor (1952); and Woodman (1969).

15. See, for example, Peter Hopkirk's *The Great Game* (1990) and John Keay's *The Gilgit Game* (1979). These texts are conspicuous on the bookshelves of most Westerners in Gilgit. They can be purchased at G. M. Baig's Bookshop in the Jama'at Khana Bazaar in Gilgit and English-language bookstores in Islamabad. Those Westerners who do not own copies of these texts tend to borrow them from friends.

16. In particular, the Lonely Planet guides to Pakistan, the KKH, and Northern Areas trekking routes are widely read among my research participants, whether they plan to trek, climb, or just get a 'feel' for the areas in which they work.

17. Indigenous men have supplied critical transport labor over the centuries to feudal rulers, Dogra rajputs, and European soldiers and adventurers who needed supplies moved through the mountains (see MacDonald 1998). That tradition continues as porters carry the luggage of contemporary metropolitan adventure tourists to off-road mountains and valleys.

18. Deborah Bhattacharyya (1997) notes a similar problem with Lonely Planet Guides to India. While authors warn women travelers about potential sexual harassment from Indian men and suggest appropriate actions to avoid such unwanted attention, they neglect to caution women against fellow foreign tourists. As do Pakistan guidebooks, these texts mediate Western women's experiences of the country in which they are traveling, as well as their representations of and relationships with indigenous men.

19. Islamization movements worldwide can be understood as acts of resistance, partly against Euro-American colonialism, cultural imperialism (including 'women's freedom,' which Islamists commonly understand as the basis of the 'West's' moral disintegration), and certain aspects of modernization (Abu-Odeh 1992; Ali 1992; Gardezi 1985; Haeri 1995; Kepel 1992; Sahgal and Yuval-Davis 1992; Tessler and Jesse 1996; Yazbeck

Haddad and Smith 1996). These social, economic, political, and cultural forces have marginalized many Muslim people, and have thus engendered anti-Western sentiments, often expressed in the form of oppositional national identities. In Pakistan, for example, fundamentalist parties such as the *Jamaat-I-Islami*, in conjunction with military state leaders, have attempted to 'purify' the nation of 'Western' influences by reasserting a distinctive Muslim identity. This identity is achieved primarily through the social, sexual, and spatial segregation of women, because women's behavior is considered the touchstone of what it means to be Muslim (see Cook 2001).

20. See Butz (1998) for a contemporary example of discursive citationality and its powerful effects on some indigenous people of northern Pakistan.

21. See Butz (1995b) for a sympathetic review of Said's work and that of his critics.

Chapter Two   Vulnerable and Spatializing Subjects

1. The Australian Institute of Criminology (Carcach and Mukherjee 1999) and the American National Survey on Crime Prevention (2001) found that the majority of women surveyed feel constantly vulnerable and that fear of crime affects nearly all women in some way. For example, approximately 70 percent of Australian women feel unsafe when walking alone at night.

2. Numerous women mentioned daily incidents of sexual harassment when riding public transit in Australia, walking the streets of London, and shopping in mid Western America.

3. Western women in Gilgit, like those who travel in India (see Bhattacharyya 1997, 386), frequently participate in a one-way ethical assessment of sexual behavior. While they feel free to scrutinize indigenous men's actions, they are offended to think those men may be evaluating their behaviors in turn. This perceived 'reverse gaze' is so unnerving that most women find it tantamount to sexual harassment. And their responses to this surveillance constitute the proof of the need for it.

4. When Andy situates Gilgiti men's behaviors in a Victorian cultural milieu, she evokes two Orientalist discourses about time. First, local history is foresworn by positioning social change in relation to European historical events and eras (Said 1978, 86). This imperialist "discourse of negation" (Gregory 1995, 36) relegates 'ineffectual' Gilgitis to historical backwardness, leaving them to rely on the Europeans for social transformation (see Fabian 1983). Second, the discourse of negation intersects with the presumption that local social conditions, as well as being unequivocally different from those in the West, remain uniform over time (Said 1978, 96). According to Andy, Gilgit is a place where nothing ever changes, especially

for Muslim women (see also my discussion of Kathleen Jamie's travel writing in chapter one). If local culture is ostensibly stuck in time, moored in a primitive past when the oppression of women was the norm, then Western women feel 'out of time,' as well as 'out of place,' in contemporary Gilgit. In contrast, they believe Muslim women are 'out of time' in the global context of women's emancipation, realized through the women's liberation movement. As I describe in chapter three, my research participants, who understand themselves as modern liberated women, hope to serve as emancipatory examples for past-bound indigenous women, to act as contemporary suffragettes who work at a small scale to free Gilgiti women, and thus return them to history.

5. Norbert Elias (1978) argues that these 'civilized' characteristics and behaviors were important in forming bourgeois notions of individuality and culture in the West. 'Cultured' German intelligentsia elaborated them in the eighteenth century to distinguish themselves from members of the ruling aristocracy. In order to achieve upward social mobility in a situation where ancestry grounded the social hierarchy, these intellectuals developed a set of bourgeois manners that emphasized depth of feeling, aspiration, virtue, and intrinsic worth—as opposed to the superficial ceremony and formal conversation typical of courtly life—as the markers of civilization. As personal depth and inner intellectual enrichment were cultivated at the expense of superficial artifice, German intellectuals, like Goethe, became 'individuals.' Elias suggests that the individual was born through this 'civilizing' process. The development of individual personality in the bourgeoisie through changes in behavior is thus an important element in explaining sociogenesis in this period of European history. I return to current incarnations of the 'cult of individuality' in Western women's traveling adventures in chapter three.

6. Anguita's comments recall those of some British women in colonial India who criticized authors such as E. M. Forster and Paul Scott for their unduly strong emphasis on white women's fear of rape by indigenous men. In their attempts to dismiss women's vulnerability to assault, Margaret Ramsay-Brown claimed that rape "never entered one's thinking," while Mrs. Symington maintained "Indians were always very courteous, even all through the noncooperation movement" (as quoted in Procida, 2002, 194–195). According to Mary Procida (2002, 195), a significant number of women who were married to Raj civil servants and military officers "did not fret about their own objectification as targets of Indian men's lust and desire," but rather about their mutual struggle for power "in the combative environment of imperial politics."

7. This characterization of Muslim men having sex with animals and other men echoes colonial-era Orientalist discourses that depict 'Oriental' men as having an innately perverse sexuality (see Clancy-Smith 1998). Their homoerotic perversity is ostensibly encouraged by Muslim social institutions such as purdah, which keeps women secluded and sexually unavailable and men sexually frustrated. The harem or polygamous marriage, on the other hand, symbolizes Muslim men's exaggerated sexuality.

8. Sarah Kofman (1982) theorizes and traces the legacy of the notion of men's 'respect' for women, especially as it is conceived and dispersed through the philosophies of Kant and Rousseau. She suggests that men's respect for women transforms the feminine object of their desiring gaze from a 'hooker' to a more socially elevated and distant entity through an optics of fascination that avoids "immediate and close relationships with [women]" (1982, 42). The idea that men's respect implies a distanced, courteous fascination with women, rather than an intimate corporeal desire for them, could help explain why my research participants are so keen to secure Gilgiti men's respect. Respect becomes yet another avenue to women's personal safety.

9. Western women's efforts to control sexuality, whether at home or abroad, is a profound marker of bourgeois morality, a discursive continuity from Victorian times (see McClintock 1995; Stoler 1989, 1995; Young 1995).

10. Mary Louise Pratt (1992, 6, 4) defines this zone as "the space of colonial encounter" "where disparate cultures meet, clash and grapple with each other, often in highly asymmetrical relations of domination and subordination."

11. See Dittman (1997) for a detailed discussion of the ethnoreligious composition of various Gilgit bazaars.

12. This preference for Ismaili Muslims over Sunnis and Shias can also be found in many colonial-era travel accounts of what is now Northern Pakistan. For instance, E. Knight (1893, 493–494) contends that Ismailis have a more "jovial character" than other types of Muslims, and John Biddulph (1880, 30) claims that Ismailis are not as religiously fanatical as Shias and Sunnis: "Mohammedanism sits but loosely on the Hunza people." Emily Lorimer (1939, 277) prefers Ismailis, much like contemporary Western women in Gilgit do, because women are not made to veil or hide themselves from nonfamilial men.

13. Moreover, because my research participants work with Ismaili men and train them to be teachers, they have considerable control over interactions with these men, which in turn lessens their fear of assault.

14. See Dea Birkett's (1992) description of how British nurses in colonial West Africa distanced themselves from the seemingly lascivious landscape by creating 'desexualized' bodies and spaces removed from the local setting.

15. Christian women may want to reestablish their faith and religious identity in Muslim space, but this need is constituted as much by their sympathy and respect for the devoutness of many Muslims as by their feelings of cultural superiority, especially Christianity's ostensible sensitivity to the equality of the gender groups. For example, René, with only a slight degree of Christian condescension, says that "I can't say anything about anyone's *heart* or mind, but from *outward* appearances [Muslims] take their religion very seriously. I mean, you see *hundreds* of people daily just praying on the street. They stop on the sidewalk, and they lay out their cloth and they say their prayers. [I respect] the *outwardly* visible practice of religion here." Christian women's ambivalence is thus also fostered by

their selective sympathy for Muslims, in conjunction with their criticisms of what they perceive to be a culturally inferior religion that condones the uncivilized treatment of women.

16. Although some people bring locally produced hashish to parties, it is not popular with most Westerners unless no other intoxicants are available. Alcohol, whether in the form of homemade wine and beer or spirits purchased in Islamabad, is the drug of choice. This group preference for expensive and scarce alcohol over cheap local drugs is not only about who does or does not inhale. It can also be understood as a ritual of cultural identity and belonging that can only be performed in Western spaces where Pakistan's Islamic law against alcohol consumption is not usually enforced.

Chapter Three 'Free' Travelers and Developers Navigating Boundaries

1. Although Western development workers in Gilgit lump tourists and travelers into a single category of 'exploitative visitor' searching for instant self-gratification, the specious distinction between 'traveler' and 'tourist' is guarded by the tourist industry to serve its own commercial ends and cultural capital needs, and by individual travelers who, by dissociating themselves from tourists, convince themselves that they are contributing to the well-being of the cultures through which they travel, rather than exploiting them (see Buzard 1993a; Holland and Huggan 1998; and MacCannell 1989, 1992).

2. James Clifford (1997, 75) describes "cultural discretion" as a visitor's act of "looking beyond 'mere conventions' or going along with appearances to a deeper level of respect based on historical knowledge and cultural comprehension."

3. Carrie Chapman Catt provides an interesting contrast to most Western women who were uninterested in the lives of colonized Muslim women. As a liberal feminist traveling under the auspices of the International Alliance of Women in 1911, she was convinced that women's universal oppression would be abolished with suffrage. This conviction "prompted a sincere eagerness to meet Muslim women and find out about their lives" (Weber 2001, 134).

4. 'Push' factors such as unemployment at home illustrate the arguments made by Holland and Huggan (1998, 5) and Van den Abeele (1992, Introduction) that the conditions of 'home' are the primary frame of reference for most Western travelers.

5. Rebecca Saunders (1999) and Nupur Chaudhuri (2002) describe some of the personal reasons British women mentioned for traveling to colonial India, which overlap with the factors that pushed my research participants from home: the desire to escape the triviality, routineness, problems, and

constraints of life at home; to seek adventure, challenging work, and an economically secure life; to feed exotic fantasies; to be more like men; and to feel important and worthwhile. British nurses in colonial West Africa had similar reasons for traveling (Birkett 1992). Across time, privileged white women use travel as a means to a freer life. Also in keeping with my analysis, Saunders discusses how these personal motives conflict with official, selfless 'civilizing' motives: metropolitan women, both in the past and present, emotionally exploit their supposedly altruistic experiences in South Asia.

6. See Pratt (1986); Crapanzano (1986, 69–70); and MacCannell (1989) for discussions of the contrasting tropes that characterize travel and ethnographic stories of arrival in the tourism/research 'field,' which I have attempted to disrupt in the introduction of this ethnography.

7. Erik Cohen (1979) designates this style of touristic experience as the 'Diversionary Mode' of travel, which is animated by a tourist's desire to escape the monotony of everyday life at home.

8. See Cocker (1992) and Holland and Huggan (1998) for discussions of travel as a door to freedom for Western tourists.

9. Abbie's link between indigenous women's 'improvement' and companionate marriage has had powerful longevity since the eighteenth century, when this type of familial relationship, with its strict gendered divisions of labor, was understood as the cornerstone of civilization. As Catherine Hall (2004, 51) argues, the attempt to cultivate "this superior form of the family, with its elevated yet inferior female, was critical to the British legitimation of colonial rule."

10. This predilection to disregard gendered commonalities differs somewhat from colonial times, when some British women in India, particularly self-proclaimed feminists, perceived a bond between themselves and indigenous women based on connections of 'sisterhood' and motherhood (Burton 1990). While contemporary 'postmodern' feminists question the notion of universal 'sisterhood,' they are more sensitive than my research participants to the subjugating effects of various discourses of gender on women around the world.

11. Like many other Western women, Louise sees herself as a resistant example for downtrodden indigenous women. However, in doing so she ignores a sign that they often do not need such an example. Faridah's 'explosion' shows she is aware of the oppression enacted on her by men of her household. Louise, therefore, overlooks the commonality of women's oppression and resistance. This story also reveals how indigenous men negotiate women's resistance by invoking a cultural relativism that delegitimates supposedly Western feminist ideas and values.

12. This widespread indifference to learning a local language is somewhat reminiscent of colonial times, when few British women living in India could communicate with indigenous people (Barr 1976; Burton 1990; Jolly 1993; MacMillan 1988; Prodica 2002; Spivak 1985; Ware 1992).

Transience and inadequate time and ability may be partly responsible for VSOs' lack of interest. But that cannot explain why most Western women in colonial India learned only a few words of a foreign language over decades of living in the country, against the advice of servant management guides.

13. See, for example, Heath (1992); March (1983); Messick (1987); Schevill et al. (1991); Schneider (1980, 1987); Weiner and Schneider (1989); but see Crane (2000); Kaiser et al. (1993); and Tarlo (1996) as notable exceptions.

14. While Western clothes enable women to negotiate experiences of belonging/not belonging in Gilgit, they also help mark the inside/outside of home, as Margaret and Louise describe. As they do in the West, my research participants often change clothes once they return home from work, replacing their 'public' clothing presentation style with more casual 'private' attire.

15. Perhaps the image of biblical women is also invoked for these women when they wear head-coverings.

16. Timothy Brennan (2001) points out that, despite the popular connotation of cosmopolitanism as a proper mode of social conduct in a global age that opposes nativism, prejudice, sameness, and closed-mindedness and promotes cross-cultural universals, in practice the term is often used to erase difference in favor of the imperialistic spread of universals derived from the West. When most of my research participants use the term cosmopolitanism to describe the outcome of their development work, they imply a global mode of conduct, but practice the imperialistic spread of universals. A few women, however, develop critiques of the latter tendency, as I will show.

17. For example, in July, 1985, the National Geographic Society joined with three other professional geographical associations to form the Geographic Education National Implementation Project (GENIP) (see http://genip. tamu.edu). Since that time, GENIP has made efforts to improve the quality of geographical education in publicly funded American school curricula from grades K-12. The initiative began as a response to most American students' inability to locate European countries on a world map.

18. Mary Ann Lind (1988) analyzes the social reform and welfare activities of 15 British women in colonial India, who, as wives of prominent civil servants and military officers, seriously engaged the government's hope that women's cross-cultural contact would enlighten Indian women to British ways of life. However, she notes that outside her sample, "only a small minority of British women came directly into contact with Indian women" (34). Mary Procida (2002, 165–173) comes to the same conclusion, arguing that most British women of this group refused the white women's burden due to their beliefs that they shared no common ground with Indian women and that philanthropy was not their primary function in the colony. Moreover, the notion that Indian women were dirty and diseased made grassroots charity unsavory. Increasingly, the government recruited

'professional' teachers, nurses, and social workers from Britain, who came for solely philanthropic reasons. Many of these women, including Maisey Wright (1988), attempted to integrate themselves more fully into indigenous communities as part of their efforts to improve the lives of Indian women.

Chapter Four   Another Bun in the Oven

1. Alison Blunt (1999a) praises the attempts made by Chaudhuri (1988); Grewal (1996); Jayawardena (1995); and Sharpe (1993) to examine how British women exercised power in colonial India, but criticizes these efforts for relying on a distinction between public and private space. In separating domestic and colonial power, these authors obscure the colonial context of domesticity and ignore the nexus of colonial power relations within British homes in India (see also Procida 2002). I bypass this problem by examining the complex regime of power relations that shape domesticity in Western homes in Gilgit.
2. This domestic solace is invoked through the Victorian image of home, which depicts the domestic sphere as a place of peace, a shelter from doubt and terror, a sacred haven of love and hospitality, and a refuge from public sphere activities and problems. See Procida (2002) and Sarup (1994).
3. My research participants are not the only transnational group of people who feel compelled to make specific kinds of bread away from home. Many racialized women who immigrate to Canada, including those from South Asia, Algeria, and Ethiopia, bring their bread-making customs and appliances to their host country as a way both to connect them to the place and people they left behind and to perpetuate their cultural traditions. Therefore, baking 'different' kinds of bread is not a practice exclusive to white middle-class women. However, I think the primary difference between white women in Gilgit and immigrant women to Canada is the formers' emphasis on the 'innately inferior' quality of many other breads, as well as the sense of cultural continuity and belonging both groups seek.
4. The connection between the space of home and the formation of identity has also been convincingly analyzed by other authors. See, in particular, Blunt (1999a) and Sarup (1994).
5. See also Foucault (1987, 1988). Governmentality refers to "all endeavors to shape, guide and direct the conduct of [self and] others" (Rose 1999, 3–4).
6. See Georgina Gowan's (2001) exploration of autobiographical accounts describing colonial women's experiences of repatriation to Britain for an examination of the tensions involved in returning to a place called 'home.'
7. Domesticity, therefore, is another mechanism through which my research participants negotiate both their perceived gender oppression and their lack of competence in Gilgit's public sphere.

8. Mary Procida (2002, 63) outlines the dominant and contrasting representations of expatriate and indigenous homes in colonial India. Closed, cramped, uncomfortable, and squalid Indian residences, the "breeding grounds for moral and physical corruption," were distinguished from the open, cozy, and sanitary expatriate living conditions that allotted the 'proper' amount of household space to each occupant.

9. In colonial India, when household décor items were largely unavailable and declining expatriate salaries made what was at hand prohibitively expensive, British homes were often meagerly outfitted with rented furniture and white-washed walls. Women indulged their decorating fantasies mainly by leafing through imported catalogues (Procida 2002). Home decorating became a more important site of the domestic management of imperialism in late-twentieth-century Gilgit as more household wares were available in southern Pakistani cities and long-term expatriates were paid international wages in nonlocal currencies.

10. Western women's household gardens also signify a desire for self-sufficiency and efficacy in a social environment where they often feel incapable and dependent on others for services and practical advice about how to live successfully in Gilgit. As Evelyn mentions, "I was really happy this morning. I made pancakes with peaches on them, and I went out to pick the peaches off my own trees to make them. There was something really self-satisfying about that independence, although I'm not sure I understand why I felt that." Moreover, home-grown fruits and vegetables allow my research participants to escape trips to the bazaar more often.

11. A properly clean house is also a metaphor for control over space, which cannot be substantially achieved in the 'outside' or public sphere of Gilgit. However, this spatial ordering practice can be enacted within the confines of home (see Douglas 1966).

12. Lines of civility are not only drawn between bourgeois Westerners and unrefined Pakistanis. Elite urban Pakistanis often differentiate themselves from *jangli* villagers based on their access to privacy, apportioned domestic space, and use of birth control. And many times they are as vigilant as Western women in Gilgit to keep their well-appointed homes free of dirt and full of a wide range of foodstuffs. Consequently, exercising discourses of class and location, Westerners often make a distinction between 'refined' city-dwelling Pakistanis and 'gauche' rural villagers.

13. The predominant form of Western hospitality—inviting people to enjoy a home-cooked meal—hardly differs from the primary local form. However, indigenous performances of hospitality are often not understood by Western women as cultivated, most often, I believe, because of the type of food served, but also because many Gilgiti women are heard but not seen at the dinner table due to purdah restrictions. Most of my research participants feel uncomfortable eating with men in a sexually segregated domestic setting. Some imagine oppressed Muslim women slaving in the kitchen and eating rummaged leftovers.

14. To recap, I use this term to describe Western women's struggles to maintain identity and self-control by cultivating their distinction from Others via gender, class, and racially coded rules for living.

15. Many VSOs could afford to hire servants if they chose to channel their money there rather than toward travel, household amenities, and imported food, because servants receive exceedingly low wages. For example, Lyn and Marion considered hiring a Gilgiti woman to clean, but decided against it once they perceived Gilgit to be a place, unlike Vietnam and Africa, where employing servants is not common. Money is apparently not the issue: "If we earn Rs. 7000 [US$115] a month and we can live on Rs. 5000 [US$82], then I'd rather that a servant earned that Rs. 2000 [US$33]. I don't want to give it away, but I would rather that someone else earned it. But it's still not part of the culture here." Despite their perception that employing servants is an unconventional indigenous cultural practice, Lyn and Marion have hired a man to wash their floors now and again.

16. Evelyn recognizes the difference servants make to her lifestyle: "I told you about those two cyclists we had to dinner who were going to Lhasa, we just met them for one night . . . I know how fortunate I am here, compared to at home. I have someone to clean my house and cook so if I have someone interesting over I can sit down and talk to them. I feel fortunate that I have that kind of lifestyle here."

17. MacMillan (1988); McClintock (1995, 258–295); Procida (2002); and Tranberg Hansen (1989) demonstrate how Western women in colonial Africa and India were also reliant on indigenous labor to shoulder large quantities of domestic work. As employers, Western women's middle-class identity and homes, today as in the past, thus take formative shape around the abjected labor of working-class indigenous people who, though repudiated, are nevertheless indispensable. See also Pollock (1994) for a tangential analysis of the intimate relationship between South African women and their black nannies as the predominant class system at the heart of the white African household.

18. Ayah is the Urdu word for nanny.

19. Flora Annie Steel and Grace Gardiner, in *The Complete Indian Housekeeper and Cook* (1905), advised British women in India not to engage with the minutiae of household management. Rather, as they inspected work, issued orders, and instituted an efficient housekeeping regime, their symbolic authority and management role would be established, leaving servants to attend to the day-to-day details of chores. Free from the everyday grind, women could undertake more amusing and fulfilling activities, including working for the improvement of indigenous people.

20. See King (1990) for a discussion of the role American women missionaries played in support of Euro-American colonialism in China. By employing Chinese servants, these women aimed to achieve cultural and religious conversions through their domestic example. King argues that American

women's role in the domestic sphere thus added structure and emotional force to other aspects of colonialism in China.

21. See Parry (1976) for an analysis of what are now termed Orientalist discourses that juxtapose a supposedly base and solely physical 'Eastern' love in arranged marriages with an ostensibly virtuous 'Western' love that engenders the finer feelings of protection and care with no thought of reward.

22. Christian missionaries in nineteenth century colonial India also assumed that indigenous people could be improved, civilized, and redeemed. However, Hall (2004, 59) argues that by the end of the century this discourse of uplift shifted as the notion of fixed racial types and aptitudes gained scientific authority.

23. As in the colonial era, rudimentary living arrangements for servants seem to be "part of the natural order of things" (Procida 2002, 91).

24. But in colonial India, some British women cultivated sexual intimacy with their servants, exploiting their imperial position to coerce sexual services from male servants (Sinha 1992).

25. In this way VSOs resemble many of the British women who arrived in India after World War I. This group of expatriates came from a broader class background than was previously typical, and were often embarrassed and discomforted by both their newfound class authority and ostentatious displays of British wealth and power in India (Procida 2002, 94).

26. Nupur Chaudhuri (1988) enumerates the criteria used by Western mothers in colonial India when hiring local ayahs. As Western women in contemporary Gilgit do, memsahibs preferred to employ English-speaking nannies who behaved more like European women in order to protect their children from the cultural 'perversions' enacted through local languages and women's spatial and social segregation.

27. According to Fiona Paisley (2004), the rhetoric of dysgenic subjects necessitated a moral education for white children in imperial colonies that taught racial responsibility through good health and hygiene, happiness, and spiritual, intellectual, and physical development. Mothers who instituted this moral regime through organized play, physical and leadership training, and mental stimulation would be certain to raise fit children for the imperial nation.

28. Chaudhuri (1988) and MacMillan (1988) discuss how British memsahibs in colonial India also sent their children away to school, usually overseas, to guarantee their cultural vigor despite the pain and worry this caused their mothers.

29. See David Matless (1995), who discusses how an appropriate sense of personhood and citizenship was forged in the first four decades of the twentieth century in England through the physical fitness routines of communist ramblers, cyclists, guides and scouts, health advocates, charabancers, and modern dancers.

30. René and her spouse had mission postings in Africa and southern Pakistan before they moved to Gilgit. In each case they planned their

overseas mission stints to begin after a child was 'safely' born in the United States.

Conclusion Ruptures and Recuperations?

1. See Clifford (1983, 1990); Clifford and Marcus (1986); Crapanzano (1986); Marcus and Fischer (1986); Said (1989); Stacey (1991); Tyler (1986); Van Maanen (1990); and Visweswaran (1994) for critiques of the violence done through traditional ethnographic representations and recommendations for alternative writing styles, including polyphonic techniques, that help avoid these difficulties.
2. Many Gilgitis, like their urban Pakistani counterparts, are also becoming global citizens as they travel to Gulf states and Iran for employment and religious pilgrimages, take cultural study tours in neighboring countries to regain lost traditions and languages, study development theory in Canada and Britain through AKF funding, and train as tour guides in Singapore, New York, and Japan.
3. For a critique of this perspective, see Weyland (1997).
4. These negotiations of gender oppression demonstrate that acts of resistance can have reactionary, as well as transformative, effects.
5. Foucault (1977c) has also written extensively about transgression as boundary crossings that illuminate discursive limits and make clear what is excluded by discourse to create new knowledge. He, like Cresswell, conceives of transgression not as a rebellion that seeks to destroy boundaries, but as an act of violence that reveals the parameters and prohibitions of limits. They thus agree that limits, in their ability to exclude and their vulnerability to exposure as limited, are constraining as well as enabling. Despite these overlaps, I follow Cresswell's formulation more closely, because it is explicitly spatial, clearer about the incidental nature of boundary transgressions and thus the type of agency involved, and more easily applied to everyday social practices.
6. Western women development workers in Gilgit are also transgressive vis à vis their own culture by breaching transnational boundaries between 'West' and 'East' to become something like cultural outlaws abroad. While this boundary challenge is especially interesting as an intrinsic feature of all 'improving' international development work, it is beyond the scope and scale of my analysis.

References

Abu-Odeh, L. 1992. Post-colonial Feminism and the Veil: Considering the Differences. *New England Law Review* 26(4): 1527–1537.

Al-Azmeh, A. 1993. *Islams and Modernities*. London: Verso.

Albrow, M. 1997. Traveling Beyond Local Cultures. In *Living the Global City*, edited by J. Eade, 37–55. New York: Routledge.

Alexander, J. 1994. *Fin de Siècle Social Theory: Relativism, Reductionism and the Problem of Reason*. London: Verso.

Ali, Y. 1992. Muslim Women and Politics of Ethnicity and Culture in Northern England. In *Refusing Holy Orders: Women and Fundamentalism in Britain*, edited by G. Sahgal and N. Yuval-Davis, 102–123. London: Virago Press Ltd.

Allan, N. 1985. Periodic and Daily Markets in Highland-Lowland Interaction Systems: Hindukush-Himalaya. In *Integrated Mountain Development*, edited by T. V. Singh and J. Kaur, 239–256. New Delhi: Himalayan Books.

Allen, C. 1976. *Plain Tales of the Raj: Images of British India in the Twentieth-Century*. New York: St. Martin's Press.

American National Survey on Crime Prevention. 2001. *Are We Safe?* Washington, DC.

Appadurai, A. 1990. Disjuncture and Difference in the Global Cultural Economy. *Public Culture* 2(2): 1–24.

Armstrong, P. and H. Armstrong. 1992. Sex and the Professions in Canada. *Journal of Canadian Studies* 27(1): 118–133.

Band, G. 1955. *Road to Rakaposhi*. London: The Travel Book Club.

Barnes, T. J. and J. S. Duncan, eds. 1992. *Writing Worlds: Discourse, Text, and Metaphor in the Representation of Landscape*. London: Routledge.

Barnett, C. 1997. 'Sing Along with the Common People': Politics, Postcolonialism and Other Figures. *Environment and Planning D: Society and Space* 15: 137–154.

Barr, P. 1976. *The Memsahibs: The Women of Victorian India*. London: Secker and Warburg.

Barthes, R. 1975. *S/Z*. London: Jonathan Cape.

Beck, U. 1992. *Risk Society: Towards a New Modernity*. London: Sage.

Bellew, H. W. 1875. *Kashmir and Kashgar: A Narrative of the Journey of the Embassy to Kashgar in 1873–74*. London: Trubner and Company.

Bhattacharyya, D. 1997. Mediating India: An Analysis of a Guidebook. *Annals of Tourism Research* 24(2): 371–389.

Biddulph, J. 1971 [1880]. *Tribes of the Hindoo Kush*. Graz, Austria: Akademische Druch-u.Verlagsanstalt.

Bielby, D. and W. Bielby. 1988. Women's and Men's Commitment to Paid Work and Family: Theories, Models, and Hypotheses. *Women's Work* 3: 249–263.

Birkett, D. 1992. The "White Women's Burden" in the "White Man's Grave": The Introduction of British Nurses in Colonial West Africa. In *Western Women and Imperialism: Complicity and Resistance*, edited by N. Chaudhuri and M. Strobel, 177–188. Indianapolis: Indiana University Press.

Blunt, A. 1994a. *Travel, Gender and Imperialism: Mary Kingsley and West Africa*. New York: The Guilford Press.

———. 1994b. Mapping Authorship and Authority: Reading Mary Kingsley's Landscape Descriptions. In *Writing Women and Space*, edited by A. Blunt and G. Rose, 51–72. New York: The Guilford Press.

———. 1999a. Imperial Geographies of Home: British Domesticity in India, 1886–1925. *Transactions of the Institute of British Geographers NS* 24: 421–440.

———. 1999b. The Flight from Lucknow: British Women Traveling and Writing Home, 1857–8. In *Writes of Passage: Reading Travel Writing*, edited by J. Duncan and D. Gregory, 92–113. London: Routledge.

Bourdieu, P. 1977. *Outline of a Theory of Practice*. Trans. R. Nice. Cambridge: Cambridge University Press.

———. 1990. *The Logic of Practice*. Trans. R. Nice. Cambridge: Cambridge University Press.

———. 1999. *The Weight of the World: Social Suffering in Contemporary Society*. Trans. P. Parkhurst Ferguson. Stanford: Stanford University Press.

———. 2001. *Masculine Domination*. Trans. R. Nice. Stanford: Stanford University Press.

Bourdieu, P. and L. Wacquant. 1992. *An Invitation to Reflexive Sociology*. Chicago: University of Chicago Press.

Boyle, J. 2000. Education for Teachers of English in China. *Journal of Education for Teaching* 26(2): 147–155.

Boyne, R. 1991. Power-Knowledge and Social Theory: The Systematic Misrepresentation of Contemporary French Social Theory in the Work of Anthony Giddens. In *Giddens' Theory of Structuration: A Critical Appreciation*, edited by C. Bryant and D. Jary, 52–73. London: Routledge.

Brah, A. 1996. *Cartographies of Diaspora: Contesting Identities*. London: Routledge.

Brantlinger, P. 1988. *Rule of Darkness: British Literature and Imperialism, 1830–1914*. Ithaca: Cornell University Press.

Breckenridge, C. and P. van der Veer. 1993. *Orientalism and the Postcolonial Predicament*. Philadelphia: University of Pennsylvania Press.

Breman, J. 1989. *Taming the Coolie Beast: Plantation Society and the Colonial Order in Southeast Asia*. Delhi: Oxford University Press.

Brennan, T. 2001. Cosmo-Theory. *South Atlantic Quarterly* 100(3): 659–691.

Bryant, C. and D. Jary. 1991. Introduction: Coming to Terms with Anthony Giddens. In *Giddens' Theory of Structuration: A Critical Appreciation*, edited by C. Bryant and D. Jary, 1–31. London: Routledge.

Bulbeck, C. 1991. New Histories of the Memsahib and Missus: The Case of Papua New Guinea. *Journal of Women's History* 3: 82–105.

————. 1998. *Re-orienting Western Feminisms: Women's Diversity in a Postcolonial World*. Cambridge: Cambridge University Press.

Burton, A. 1990. The White Woman's Burden: British Feminists and the Indian Woman, 1865–1915. *Women's Studies International Forum* 13(4): 295–308.

Bush, B. 2004. Gender and Empire: The Twentieth Century. In *Gender and Empire*, edited by P. Levine, 77–111. Oxford: Oxford University Press.

Butler, J. 1990. *Gender Trouble: Feminism and the Subversion of Identity*. London: Routledge.

————. 1993. *Bodies That Matter: On the Discursive Limits of "Sex."* New York: Routledge.

————. 2004. *Undoing Gender*. London: Routledge.

Butz, D. 1993. *Developing Sustainable Communities: Community Development and Modernity in Shimshal, Pakistan*. Unpublished Ph.D. dissertation. McMaster University, Hamilton.

————. 1995a. Legitimating Porter Regulation in an Indigenous Community in Northern Pakistan. *Environment and Planning D: Society and Space* 13(4): 381–414.

————. 1995b. Revisiting Edward Said's *Orientalism*. *Brock Review* 4(1/2): 54–80.

————. 1998. Orientalist Representations of Resource Use in Shimshal, Pakistan, and Their Extra-Discursive Effects. In *Karakorum-Hindukush-Himalaya: Dynamics of Change. Part II*, edited by I. Stellrecht, 357–386. Köln: Rüdiger Köppe Verlag.

————. 2002. Sustainable Tourism and Everyday Life in Shimshal, Pakistan. *Tourism Recreation Research* 27(3): 53–65.

Buzard, J. 1993a. *The Beaten Track: European Tourism, Literature and the Ways to "Culture" 1800–1918*. Oxford: Clarendon Press.

————. 1993b. Victorian Women and the Implications of Empire. *Victorian Studies* 36: 443–453.

Caesar, J. 1991. Slaves in the Gulf. *Progressive* 55(11): 37.

Callaway, H. 1987. *Gender, Culture and Empire: European Women in Colonial Nigeria*. Chicago: University of Illinois Press.

Canadian Briefing Center. 1986. *Pakistan: Post Report*. Quebec: Canadian International Development Agency.

Carcach, C. and S. Mukherjee. 1999. Women's Fear of Violence in the Community. *Australian Institute of Criminology* 135: 1–6.

Certeau, M. 1984. *The Practice of Everyday Life*. Trans. S. Rendall. Berkeley: University of California Press.

Chaudhuri, N. 1988. Memsahibs and Motherhood in Nineteenth-Century India. *Victorian Studies* 31(4): 517–536.

————. 1992. Shawls, Jewelry, Curry, and Rice in Victorian Britain. In *Western Women and Imperialism: Complicity and Resistance*, edited by N. Chaudhuri and M. Strobel, 231–246. Indianapolis: Indiana University Press.

Chaudhuri, N. 1994. Memsahibs and Their Servants in Nineteenth-Century India. *Women's History Review* 3(4): 549–562.

———. 2000. Issues of Race, Gender and Nation in *Englishwomen's Domestic Magazine* and *Queen*, 1850s–1900. In *Negotiating India in the Nineteenth-Century Media*, edited by D. Finkelstein and D. Peers, 51–62. New York: Palgrave Macmillan.

———. 2002. The Indian Other: Reactions to Two Anglo-Indian Women Travel Writers, Eliza Fay and A.U. In *Women and the Colonial Gaze*, edited by T. Hunt and M. Lessard, 125–134. New York: New York University Press.

Christie, C. 1994. British Literary Travelers in South East Asia in an Era of Colonial Retreat. *Modern Asian Studies* 28(4): 673–737.

Clancy-Smith, J. 1998. Islam, Gender, and Identities in the Making of French Algeria, 1830–1962. In *Domesticating The Empire: Race, Gender, and Family Life in French and Dutch Colonialism*, edited by J. Clancy-Smith and F. Gouda, 154–174. Charlottesville: University Press of Virginia.

Clancy-Smith, J. and F. Gouda, eds. 1998. *Domesticating the Empire: Race, Gender, and Family Life in French and Dutch Colonialism*. Charlottesville: University Press of Virginia.

Clifford, J. 1983. On Ethnographic Authority. *Representation* 1(2): 118–146.

———. 1990. Notes on (Field)notes. In *Fieldnotes: The Makings of Anthropology*, edited by R. Sanjek, 47–70. Ithaca: Cornell University Press.

———. 1997. Spatial Practices: Fieldwork, Travel and the Disciplining of Anthropology. In *Routes: Travel and Translation in the Late Twentieth Century*. Cambridge: Harvard University Press.

Clifford, J. and G. Marcus. 1986. *Writing Culture: The Poetics and Politics of Ethnography*. Berkeley: University of California Press.

Cobham, T. 1997. *Tourism and Environment in Nagar*. Gilgit: International Union for the Conservation of Nature.

Cocker, M. 1992. *Loneliness and Time: British Travel Writing in the Twentieth Century*. London: Secker and Warburg.

Cockerill, G. K. 1922. Byways of Hunza and Nagar. *The Geographical Journal* 60(2): 97–112.

Cohen, E. 1979. A Phenomenology of Tourist Experiences. *Sociology* 13: 179–201.

Conway, W. M. 1894. *Climbing and Exploration in the Karakorum-Himalayas*. New York: Appleton and Company.

Cook, N. 2001. The Discursive Constitution of Pakistani Women: The Articulation of Gender, Nation, and Islam. *Atlantis: A Women's Studies Journal* 25(2): 31–41.

Crane, D. 2000. *Fashion and Its Social Agenda: Class, Gender and Identity in Clothing*. Chicago: University of Chicago Press.

Crapanzano, V. 1986. Hermes' Dilemma: The Masking of Subversion in Ethnographic Description. In *Writing Culture: The Poetics and Politics of*

Ethnography, edited by J. Clifford and G. Marcus, 51–76. Berkeley: University of California Press.

Cresswell, T. 1996. *In Place/Out of Place: Geography, Ideology and Transgression*. Minneapolis: University of Minnesota Press.

Crompton, R. 1987. Gender, Status, and Professionalism. *Sociology* 21(3): 413–428.

Curzon, G. 1927. *Leaves from a Viceroy's Notebook and Other Papers*. London: Macmillan.

Dani, A. H. 1991. *History of the Northern Areas of Pakistan*. Islamabad: National Institute of Historical and Cultural Research.

Davies, C. 1996. The Sociology of Professions and the Profession of Gender. *Sociology* 30(4): 661–678.

Dean, M. 1994. *Critical and Effective Histories: Foucault's Methods and Historical Sociology*. London: Routledge.

de Filippi, F. 1915. Expedition to the Karakorum and Central Asia, 1913–14. *The Geographical Journal* 46(2): 85–105.

Development Research Group. 1995. *Mountain Tourism in the Northwest Frontier Province and Northern Areas of Pakistan*. Kathmandu: International Center for Integrated Mountain Development.

Diamond, I. and L. Quinby. 1988. Introduction. In *Feminism and Foucault: Reflections on Resistance*, edited by I. Diamond and L. Quinby, ix–xx. Boston: Northeastern University Press.

Dittman, A. 1997. Central Goods and Ethno-Linguistic Groups in the Bazaars of Northern Pakistan: An Example of Central Place Theory Modifications in Mountainous Environments. In *Perspectives on History and Change in the Karakorum, Hindukush and Himalaya*, edited by I. Stellrecht and M. Winiger, 118–130. Köln: Rüdiger Köppe Verlag.

Dobbs, R. 2000. Women's Fear of Crime: Is It Fear of Rape? Paper presented at the American Society of Criminology Conference. Los Angeles.

Dolphin, T. 2000. The Discursive Construction of Hunza, Pakistan in Travel Writing: 1889–1999. Unpublished M.A. Thesis. Carleton University, Ottawa.

Douglas, M. 1966. *Purity and Danger*. London: Routledge.

Drew, F. 1875. *The Jummoo and Kashmir Territories: A Geographical Account*. New Delhi: Cosmo.

Durand, A. 1899. *The Making of a Frontier: Five Years of Experiences and Adventures in Gilgit, Hunza, Nagar, Chitral, and the Eastern Hindu Kush*. London: John Murray.

Earl, B. 1930. *Trekking in Kashmir: With Family, or without One*. Lahore: Civil and Military Gazette Press.

Elias, N. 1978. *The Civilizing Process: The History of Manners*. Trans. E. Jephcott. New York: Urizen Books.

Engels, D. 1990. History and Sexuality in India: Discursive Trends. *Trends in History* 4(4): 15–42.

Escobar, A. 1995. *Encountering Development*. Princeton: Princeton University Press.

Etherton, P. T. 1911. *Across the Roof of the World: A Record of Sport and Travel through Kashmir, Gilgit, Hunza, the Pamirs, Chinese Turkestan, Mongolia, and Siberia*. London: Constable and Company.

Fabian, J. 1983. *Time and the Other: How Anthropology Makes Its Object*. New York: Columbia University Press.

Fanon, F. 1967. *Black Skin, White Masks*. Trans. C. Lam Markmann. New York: Grove Weidenfeld.

Farnell, B. 2000. Getting Out of the Habitus: An Alternative Model of Dynamically Embodied Social Action. *Journal of the Royal Anthropological Institute* 6: 397–418.

Ferguson, J. 1994. *The Anti-politics Machine: 'Development,' Depoliticization, and Bureaucratic Power in Lesotho*. Minneapolis: University of Minnesota Press.

Fisher, B. 1990. Alice in the Human Services: A Feminist Analysis of Women in the Caring Professions. In *Circles of Care: Work and Identity in Women's Lives*, edited by E. Abel and M. Nelson, 108–131. New York: State University of New York Press.

Fisher, B. and J. Sloan. 2000. Unraveling Fear of Crime among College Women. Paper presented at the American Society of Criminology Conference. Los Angeles.

Fleming, L. 1992. American Missionaries' Ideals for Women in North India, 1870–1930. In *Western Women and Imperialism: Complicity and Resistance*, edited by N. Chaudhuri and M. Strobel, 191–206. Indianapolis: Indiana University Press.

Formes, M. 1995. Beyond Complicity versus Resistance: Recent Work on Gender and European Imperialism. *Journal of Social History* 28: 629–641.

Forster, E. M. 1976 [1924]. *A Passage to India*. Kent: Harcourt International.

Fortier, A. 1999. Re-membering Places and the Performance of Belongings. *Theory, Culture and Society* 16(2): 41–64.

Foucault, M. 1972. *The Archaeology of Knowledge and the Discourse on Language*. Trans. R. Sawyer. New York: Pantheon Press.

———. 1977a. *Discipline and Punish: The Birth of the Prison*. Trans. A. Sheridan. New York: Vintage Books.

———. 1977b. Nietzche, Genealogy, History. In *Language, Counter-Memory, Practice*. Trans. D. Bouchard and S. Simon, edited by D. Bouchard, 139–164. Ithaca: Cornell University Press.

———. 1977c. A Preface to Transgression. In *Language, Counter-Memory, Practice*. Trans. D. Bouchard and S. Simon, edited by D. Bouchard, 29–52. Ithaca: Cornell University Press.

———. 1978. *The History of Sexuality, Volume 1*. Trans. R. Hurley. New York: Vintage Books.

———. 1980. *Power/Knowledge: Selected Interviews and Other Writings*. Trans. C. Gordon, L. Marshall, J. Mepham, and K. Soper. Brighton: Harvester.

———. 1984. What Is Enlightenment? In *The Foucault Reader*. Trans. C. Porter, edited by P. Rabinow, 32–50. New York: Pantheon Books.

———. 1987. The Ethic of Care for the Self as a Practice of Freedom. In *The Final Foucault*, edited by J. Bernauer and D. Rasmussen, 1–20. Cambridge, MA: MIT Press.

———. 1988. *Technologies of the Self: A Seminar with Michel Foucault*, edited by L. Martin, H. Gutman, and P. Hutton. Amherst: University of Massachusetts Press.

Fowler, M. 1987. *Below the Peacock Fan: First Ladies of the Raj*. New York: Penguin Books.

Fraser, N. and L. Nicholson. 1990. Social Criticism without Philosophy: An Encounter between Feminism and Postmodernism. In *Feminism/Postmodernism*, edited by L. Nicholson, 19–38. New York: Routledge.

Freeman, S. 1990. *Managing Lives: Corporate Women and Social Change*. Amherst: University of Massachusetts Press.

Friedberger, R. and V. Pooley. 1995. *AKRSP's Community Tourism Action Program*. Gilgit: Human Resource Development Institute.

Frye, M. 1983. *The Politics of Reality: Essays in Feminist Theory*. New York: The Crossing Press.

Fuss, D. 1989. *Essentially Speaking*. New York: Routledge.

Gardezi, H. 1985. The Postcolonial State in South Asia: The Case of Pakistan. *South Asia Bulletin* 5(2): 1–7.

Garnham, N. and W. Williams. 1990. Pierre Bourdieu and the Sociology of Culture: An Introduction. *Media, Culture and Society* 2: 209–223.

Geertz, C. 2001. Deep Hanging Out. In *Available Light: Anthropological Reflections on Philosophical Topics*. Princeton: Princeton University Press.

George, R. 1994. Homes in the Empire, Empires in the Home. *Cultural Critique* 26 (Winter): 95–127.

Ghose, I. 1998. *Women Travelers in Colonial India: The Power of the Female Gaze*. Delhi: Oxford University Press.

Giddens, A. 1979. *Central Problems in Social Theory: Action, Structure and Contradiction in Social Analysis*. London: Macmillan.

———. 1984. *The Constitution of Society*. Cambridge: Polity Press.

Gilman, S. 1986. Black Bodies, White Bodies: Toward an Iconography of Female Sexuality in Late Nineteenth-Century Art, Medicine and Literature. In *Race, Writing and Difference*, edited by H. L. Gates, 223–259. Chicago: University of Chicago Press.

Gini, A. and T. Sullivan. 1988. Women's Work: Seeking Identity through Occupations. *Employee Responsibilities and Rights Journal* 1(1): 39–45.

Goudge, P. 2003. *The Whiteness of Power: Racism in Third World Development and Aid*. London: Lawrence and Wishart.

Gowan, G. 2001. Gender, Imperialism and Domesticity: British Women Repatriated from India, 1940–1947. *Gender, Place and Culture* 8(3): 255–269.

Gratz, K. 1997. Walking on Women's Paths in Gilgit: Gendered Space, Boundaries and Boundary Crossings. In *Karakorum-Hindukush-Himalaya: Dynamics of Change. Part II*, edited by I. Stellrecht, 489–507. Köln: Rüdiger Köppe Verlag.

Gregory, D. 1995. Between the Book and the Lamp: Imaginative Geographies of Egypt, 1849–50. *Transactions of the Institute of British Geography NS* 20: 29–57.

Grewal, I. 1996. *Home and Harem: Nation, Gender, Empire, and the Cultures of Travel*. Durham: Duke University Press.

Griswold, W. 1994. *Cultures and Societies in a Changing World*. Thousand Oaks: Pine Forge Press.

Haeri, S. 1995. Of Feminism and Fundamentalism in Iran and Pakistan. *Contention* 4(3): 129–149.

Haggis, J. 1990. Gendering Colonialism or Colonizing Gender? Recent Women's Studies Approaches to White Women and the History of British Colonialism. *Women's Studies International Forum* 13: 105–115.

Hall, C. 2004. Of Gender and Empire: Reflections on the Nineteenth-Century. In *Gender and Empire*, edited by P. Levine, 46–76. Oxford: Oxford University Press.

Hannerz, U. 1993. The Cultural Roles of World Cities. In *Humanizing the City?* edited by A. Cohen and K. Fukuo, 67–84. Edinburgh: Edinburgh University Press.

———. 1996. *Transnational Connections*. London: Routledge.

Haraway, D. 1985. A Manifesto for Cyborgs: Science, Technology, and Socialist Feminism in the 1980s. *Socialist Review* 15(80): 65–107.

Hartsock, N. 1996. Postmodernism and Political Change: Issues for Feminist Theory. In *Feminist Interpretations of Michel Foucault*, edited by S. Hekman, 39–55. Pennsylvania: Pennsylvania University Press.

Harvey, D. 1996. *Justice, Nature and the Geography of Difference*. Oxford: Blackwell Publishers.

Hassan, R. 1995. Rights of Women within Islamic Countries. *Canadian Woman Studies* 15(2/3): 40–44.

Hatem, M. 1992. Through Each Other's Eyes: The Impact of the Colonial Encounter on the Images of Egyptian, Levantine-Egyptian, and European Women, 1862–1920. In *Western Women and Imperialism: Complicity and Resistance*, edited by N. Chaudhuri and M. Strobel, 35–58. Indianapolis: Indiana University Press.

Heath, D. 1992. Fashion, Anti-fashion, and Heteroglossia in Urban Senegal. *American Ethnologist* 19(1): 19–32.

Hebard, C. 1996. Managing Effectively in Asia. *Training and Development* 50 (4): 36–39.

Hekman, S., ed. 1996. *Feminist Interpretations of Foucault*. University Park: Pennsylvania State University Press.

Henrickson, J. 1960. *Holiday in Hunza*. Washington, DC: Review and Herald Publishing Association.

Heron, B. 1999. Desire for Development: The Education of White Women as Development Workers. Unpublished Ph.D. dissertation. University of Toronto, Toronto, Ontario, Canada.

Hoggett, P. 1992. A Place for Experience: A Psychoanalytic Perspective on Boundary, Identity and Culture. *Environment and Planning D* 10: 345–356.

Holland, P. and G. Huggan. 1998. *Tourists with Typewriters: Critical Reflections on Contemporary Travel Writing*. Ann Arbor: The University of Michigan Press.

hooks, b. 1984. *Feminist Theory from Margins to Center*. Boston: South End Press.

Hopkirk, P. 1990. *The Great Game: On Secret Service in High Asia*. London: Murray.

Inglis, A. 1975. *The White Women's Protection Ordinance: Sexual Anxiety and Politics in Papua*. New York: Palgrave.

Irigaray, L. 1985 [1977]. *This Sex Which Is Not One*. Trans. C. Porter and C. Burke. Ithaca: Cornell University Press.

Iturrizaga, L. 1997. Preliminary Results of Field Observations on the Typology of Post-Glacial Debris Accumulations in the Karakorum and Himalaya Mountains. In *Karakorum-Hindukush-Himalaya: Dynamics of Change. Part* II, edited by I. Stellrecht, 71–98. Köln: Rüdiger Köppe Verlag.

Jamie, K. 1993. *The Golden Peak: Travels in Northern Pakistan*. London: Virago Press.

Jayawardena, K. 1995. *The White Woman's Other Burden: Western Women and South Asia during British Rule*. London: Routledge.

Jayaweera, S. 1990. European Women Educators under the British Colonial Administration in Sri Lanka. *Women's Studies International Forum* 13(4): 323–331.

Jewitt, S. 1995. Europe's "Others"? Forestry Policy and Practices in Colonial and Postcolonial India. *Environment and Planning D: Society and Space* 13: 67–90.

Jhabvala, R. P. 1975. *Heat and Dust*. Washington, DC: Counterpoint.

Jiwani, Y. 1992. The Exotic, the Erotic, and the Dangerous: South Asian Women in Popular Film. *Canadian Woman Studies* 13(1): 42–46.

Jolly, M. 1993. The Maternal Body and Empire. In *Feminism and the Politics of Difference*, edited by S. Gunew and A. Yeatman, 103–127. Halifax: Fernwood Publishing.

Kaiser, S., C. M. Freeman, and J. L. Chandler. 1993. Favorite Clothes and Gendered Subjectivities: Multiple Readings. *Symbolic Interaction* 15: 27–50.

Kamal, P. D. and M. J. Nasir. 1997. The Impact of the Karakorum Highway on the Land Use of the Northern Areas. In *Karakorum-Hindukush-Himalaya: Dynamics of Change. Part II*, edited by I. Stellrecht, 303–317. Köln: Rüdiger Köppe Verlag.

Kandiyoti, D. 1993. Identity and Its Discontents: Women and the Nation. In *Colonial Discourse and Postcolonial Theory: A Reader*, edited by P. Williams and L. Chisman, 376–391. New York: Harvester Wheatsheaf.

Kapman, E. 1997. *Looking for the Other: Feminism, Film, and the Imperial Gaze*. London: Routledge.

Katz, C. 2001. On the Grounds of Globalization: A Topography for Feminist Political Engagement. *Signs: A Journal of Women in Culture and Society* 26(4): 1213–1234.

Kaye, M. M. 1997. *The Far Pavilions*. New York: St. Martin's Press.

Kealy, D. 1990. *Cross-Cultural Effectiveness: A Study of Canadian Technical Advisors*. Centre for Intercultural Learning: Canadian Foreign Service Institute.

Keay, J. 1979. *The Gilgit Game: The Explorers of the Western Himalayas 1865–95*. Oxford: Oxford University Press.

Keller, E. 1989. Feminism and Science. In *Women, Knowledge, and Reality: Explorations in Feminist Philosophy*, edited by A. Garry and M. Pearsall, 175–188. Boston: Unwin Hyman.

———. 1992. How Gender Matters, or, Why It's so Hard for Us to Count Past Two. In *Inventing Women: Science, Technology, and Gender*, edited by G. Kirkup and L. Keller, 42–56. Cambridge: Polity Press.

Kepel, G. 1992. ReIslamisation Movements in Contemporary History. *Contention* 2(1): 151–159.

Kilminster, R. 1991. Structuration Theory as a World-View. In *Giddens' Theory of Structuration: A Critical Appreciation*, edited by C. Bryant and D. Jary, 74–115. London: Routledge.

Kincaid, J. 1988. *A Small Place*. New York: Penguin Books.

King, A. 1995. Re-presenting World Cities: Cultural Theory/Social Practice. In *World Cities in a World System*, edited by P. Knox and P. Taylor, 215–231. Cambridge: Cambridge University Press.

King, J. 1993. *Karakorum Highway: The High Road to China*. Hawthorn, Australia: Lonely Planet Publications.

King, M. 1990. American Women's Open Door to Chinese Women: Which Way Does It Open? *Women's Studies International* 13(4): 369–379.

Knapman, C. 1986. *White Women in Fiji, 1835–1930*. London: Allen and Unwin.

Knight, E. 1971 [1893]. *Where Three Empires Meet: A Narrative of Recent Travel in Kashmir, Western Tibet, Gilgit and Adjoining Countries*. Taipei: Ch'eng Wen Publishing Company.

Kofman, S. 1982. *Le Respect des Femmes*. Paris: Galilée.

Koskela, H. 1999. Gendered Exclusions: Women's Fear of Violence and Changing Relations to Space. *Geografiska Annaler* 81b(2): 111–125.

Kreutzmann, H. 1991. The Karakorum Highway: The Impact of Road Construction on Mountain Societies. *Modern Asian Studies* 25(4): 711–736.

Kristeva, J. 1981. Women's Time. Trans. A. Jardine and H. Blake. *Signs: Journal of Women in Culture and Society* 7(1): 13–35.

———. 1982. *Powers of Horror: An Essay on Abjection*. New York: Columbia University Press.

Kruks, S. 2001. *Retrieving Experience: Subjectivity and Recognition in Feminist Politics*. Ithaca: Cornell University Press.

Lasch, C. 1977. *Haven in a Heartless World: The Family Besieged*. New York: Basic Books.

Leidner, R. 1991. Serving Hamburgers and Selling Insurance: Gender, Work and Identity in Interactive Service Jobs. *Gender and Society* 5(2): 154–177.

Leitner, G. W. 1877. *Languages and Races of Dardistan*. Lahore: Government Central Book Depot.

———. 1889. *The Hunza-Nagyr Handbook*. Lahore: Government Central Book Depot.

Lessard, M. 2002. Civilizing Women: French Colonial Perceptions of Vietnamese Womanhood and Motherhood. In *Women and the Colonial Gaze*, edited by T. Hunt and M. Lessard, 148–162. New York: New York University Press.

Levine, P. 2003. *Prostitution, Race and Politics: Policing Venereal Disease in the British Empire*. New York: Routledge.

———. 2004a. Why Gender and Empire? In *Gender and Empire*, edited by P. Levine, 1–13. Oxford: Oxford University Press.

———. 2004b. Sexuality, Gender and Empire. In *Gender and Empire*, edited by P. Levine, 134–155. Oxford: Oxford University Press.

Lewis, R. 1996. *Gendering Orientalism: Race, Femininity and Representation*. London: Routledge.

Lind, M. 1988. *The Compassionate Memsahib: Welfare Activities of British Women in India, 1900–1947*. Westport: Greenwood Press.

Locher-Scholten, E. 1998. So Close and Yet So Far: The Ambivalence of Dutch Colonial Rhetoric on Javanese Servants in Indonesia, 1900–1942. In *Domesticating Empire: Race, Gender, and Family Life in French and Dutch Colonialism*, edited by J. Clancy-Smith and F. Gouda, 131–154. Charlottesville: University Press of Virginia.

Lorde, A. 1981. An Open Letter to Mary Daly. In *This Bridge Called My Back: Writings of Radical Women of Color*, edited by C. Moraga and G. Anzaldua, 94–97. Waterdown, MA: Persephone Press.

Lorimer, E. 1939. *Language Hunting in the Karakorum*. London: George Allen and Unwin.

Lowe, L. 1991. *Critical Terrains: French and British Orientalisms*. Ithaca: Cornell University Press.

Lyotard, J. 1984. *The Postmodern Condition: A Report on Knowledge*. Trans. G. Bennington and B. Massumi. Minneapolis: University of Minnesota Press.

MacCannell, D. 1989. *The Tourist: A New Theory of the Leisure Class*. New York: Schocken.

———. 1992. *Empty Meeting Grounds: The Tourist Papers*. New York: Routledge.

MacDonald, K. 1998. Push and Shove: Spatial History and the Construction of a Portering Economy in Northern Pakistan. *Comparative Studies in Society and History* 40(2): 287–317.

MacKinnon, C. 1982. Feminism, Marxism, Method, and the State: An Agenda for Theory. *Signs: A Journal of Women in Culture and Society* 7(3): 515–544.

MacMillan, M. 1988. *Women of the Raj*. New York: Thames and Hudson.

Madge, C. 1997. Public Parks and the Geography of Fear. *Tijdschrift voor Economische en Sociale Gegrafie* 88(3): 237–250.

Maggi, W. 2001. *Our Women Are Free: Gender and Ethnicity in the Hindukush*. Ann Arbor: The University of Michigan Press.

Malkki, L. 1997. News and Culture: Transitory Phenomena and the Fieldwork Tradition. In *Anthropological Locations: Boundaries and Grounds of a Field Science*, edited by A. Gupta and J. Ferguson, 86–101. Berkeley: University of California Press.

March, K. 1983. Weaving, Writing, and Gender. *Man* 18: 729–744.

Marcus, G. and M. Fischer. 1986. *Anthropology as Cultural Critique: An Experimental Moment in the Human Sciences*. Chicago: Chicago University Press.

Marx, K. 1976. *Capital, Volume One*. Trans. B. Fowkes. New York: Penguin Books.

Mason, K. 1931. Expedition Notes: Tours in the Gilgit Agency. *Himalayan Journal* 3: 110–115.

Matless, D. 1995. The Art of Right Living: Landscape and Citizenship, 1918–39. In *Mapping the Subject: Geographies of Cultural Transformation*, edited by S. Pile and N. Thrift, 93–122. London: Routledge.

McCall, L. 1992. Does Gender Fit? Bourdieu, Feminism and Conceptions of the Social Order. *Theory and Society* 21(6): 837–867.

McClintock, A. 1995. *Imperial Leather: Race, Gender and Sexuality in the Colonial Contest*. New York: Routledge.

McCulloch, J. 2000. *Black Peril, White Virtue: Sexual Crime in Southern Rhodesia, 1902–35*. Bloomington, IN: Indiana University Press.

McDowell, L. 1997. *Capital Culture: Gender at Work in the City*. London: Blackwell Publishers.

McEwan, C. 1994. Encounters with West African Women: Textual Representations of Difference by Women Abroad. In *Writing Women and Space*, edited by A. Blunt and G. Rose, 73–100. New York: The Guilford Press.

McNay, L. 1999. Gender, Habitus and the Field: Pierre Bourdieu and the Limits of Reflexivity. *Theory, Culture and Society* 16(1): 95–117.

———. 2000. *Gender and Agency: Reconfiguring the Subject in Feminist and Social Theory*. Cambridge: Polity Press.

Melman, B. 1992. *Women's Orients: English Women and the Middle East, 1718–1918*. Ann Arbor: University of Michigan Press.

Mercer, K. and I. Julien. 1988. Race, Sexual Politics and Black Masculinity: A Dossier. In *Male Order: Unwrapping Masculinity*, edited by R. Chapman and J. Rutherford, 97–164. London: Lawrence and Wishart.

Messick, B. 1987. Subordinate Discourse: Women, Weaving, and Gender Relations in North Africa. *American Ethnologist* 14: 210–225.

Michaud, R. and S. Michaud. 1975. Trek to Lofty Hunza . . . and Beyond. *National Geographic Magazine* 148(5): 644–670.

Midgley, C. 1998. *Gender and Imperialism*. Manchester: Manchester University Press.

Mies, M. and V. Bennholdt-Thomsen. 2000. *The Subsistence Perspective: Beyond the Globalized Economy*. London: Zed Books.

Mills, S. 1991. *Discourses of Difference: An Analysis of Women's Travel Writing and Colonialism*. New York: Routledge.

———. 1994. Knowledge, Gender, and Empire. In *Writing Women and Space*, edited by A. Blunt and G. Rose, 29–51. New York: The Guilford Press.

———. 1996. Gender and Colonial Space. *Gender, Place and Culture: A Journal of Feminist Geography* 3(2): 125–142.

Mindry, D. 2001. Non-governmental Organizations, "Grassroots," and the Politics of Virtue. *Signs: Journal of Women in Culture and Society* 26(4): 1187–1211.

Minh-ha, T. 1989. *Women, Native, Other: Writing Postcoloniality and Feminism*. Indianapolis: Indiana University Press.

Mitchell, T. 1995. The Object of Development: America's Egypt. In *The Power of Development*, edited by J. Crush, 129–157. New York: Routledge.

Mock, J. and K. O'Neil. 1996a. *Trekking in the Karakorum and Hindukush*. Hawthorn, Australia: Lonely Planet Publications.

———. 1996b. *Survey in Ecotourism Potential in the Biodiversity Project Area*. Gilgit: IUCN-Pakistan.

Mohammad, R. 1999. Marginalization, Islamism, and the Production of the "Other's" "Other." *Gender, Place and Culture* 6(3): 221–240.

Mohanty, C. 1984. Under Western Eyes: Feminist Scholarship and Colonial Discourses. *Boundary 2* 12(3)/13(1): 333–353.

Morrison, M. 1904. *A Lonely Summer in Kashmir*. London: Duckworth and Company.

Murdock. J. 1997. Towards a Geography of Heterogeneous Associations. *Progress in Human Geography* 21(3): 321–337.

Murphy, D. 1977. *Where the Indus Is Young: Walking in Baltistan*. London: Arrow Books Limited.

Nadis, M. 1957. Evolution of the Sahib. *The Historian* 19(4): 430–434.

Nagar, R., V. Lawson, L. McDowell, and S. Hanson. 2002. Locating Globalization: Feminist (Re)readings of the Subjects and Spaces of Globalization. *Economic Geography* 78(3): 257–284.

Nair, J. 1990. Uncovering the Zenana: Visions of Indian Womanhood in Englishwomen's Writings, 1813–1940. *Journal of Women's History* 2(1): 8–34.

Nandy, A. 1983. *The Intimate Enemy: Loss and Recovery of Self under Colonialism*. Delhi: Oxford University Press.

Nelson, L. 1999. Bodies (and Spaces) Do Matter: The Limits of Performativity. *Gender, Place and Culture* 6(4): 331–353.

Nicholson, L., ed. 1990. *Feminism/Postmodernism*. New York: Routledge.

Oddie, G. 1994. "Orientalism" and British Protestant Missionary Constructions of India in the Nineteenth Century. *South Asia* 17(2): 27–42.

Pain, R. 1991. Space, Sexual Violence, and Social Control: Integrating Geographical and Feminist Analyses of Women's Fear of Crime. *Progress in Human Geography* 15(4): 415–431.

Pain, R. 1997. Social Geographies of Women's Fear of Crime. *Transactions: The Institute of British Geographers* 22: 231–244.

———. 2000. Place, Social Relations and the Fear of Crime: A Review. *Progress in Human Geography* 24(3): 365–387.

Paisley, F. 2004. Childhood and Race: Growing Up in the Empire. In *Gender and Empire*, edited by P. Levine, 240–259. Oxford: Oxford University Press.

Papastergiadis, N. 2000. *The Turbulence of Migration: Globalization, Deterritorialization, and Hybridity*. Cambridge: Polity Press.

Parfitt, B. 1998. *Working across Cultures: A Study of Expatriate Nurses in Developing Countries in Primary Health Care*. Aldershot: Ashgate Publishing.

Parry, B. 1976. *Delusions and Discoveries*. New York: Penguin Press.

Paxton, N. 1990. Feminism under the Raj: Complicity and Resistance in the Writings of Flora Annie Steel and Annie Besant. *Women's Studies International Forum* 13(4): 333–346.

———. 1992. Mobilizing Chivalry: Rape in British Novels about the Indian Uprising of 1857. *Victorian Studies* 36(1): 5–30.

Philo, C. 1989. "Enough to Drive One Mad": The Organization of Space in Nineteenth Century Lunatic Asylums. In *The Power of Geography: How Territories Space Social Life*, edited by J. Wolch and M. Dear, 258–290. Boston: Unwin Hyman.

Philp, M. 1985. Michel Foucault. In *The Return of Grand Theory in the Human Sciences*, edited by Q. Skinner, 65–81. Cambridge: Cambridge University Press.

Pierson, R. and N. Chaudhuri, eds. 1998. *Nation, Empire, Colony: Historicizing Gender and Race*. Bloomington, IN: Indiana University Press.

Pieterse, J. 1992. *White on Black: Images of Africa and Blacks in Western Popular Culture*. New Haven: Yale University Press.

Pile, S. 1993. Human Agency and Human Geography Revisited: A Critique of "New Models" of the Self. *Transactions of the Institute of British Geographers* 18: 122–139.

———. 1997. Opposition, Political Identities, and Spaces of Resistance. In *Geographies of Resistance*, edited by S. Pile and M. Keith, 1–32. London: Routledge.

Pinney, C. 1989. Representations of India: Normalization and the Other. *Pacific Viewpoint* 29(2): 144–162.

Pollock, G. 1994. Territories of Desire: Reconsiderations of an African Childhood: Dedicated to a Woman Whose Name Was Not Really "Julia." In *Traveler's Tales: Narratives of Home and Displacement*, edited by G. Robertson, M. Mash, L. Tickner, J. Bird, B. Curtis, and T. Putnam, 63–92. London: Routledge.

Pratt, M. 1986. Fieldwork in Common Places. In *Writing Culture: The Poetics and Politics of Ethnography*, edited by J. Clifford and G. Marcus, 27–50. Berkeley: University of California Press.

————. 1992. *Imperial Eyes: Travel Writing and Transculturation*. London: Routledge.

Procida, M. 2002. *Married to the Empire: Gender, Politics and Imperialism in India, 1883–1947*. Manchester: Manchester University Press.

Rafael, V. 1995. Colonial Domesticity: White Women and United States Rule in the Philippines. *American Literature* 67: 639–666.

Ramusack, B. 1990. Cultural Missionaries, Maternal Imperialists, Feminist Allies: British Women Activists in India, 1865–1945. *Women's Studies International Forum* 13(4): 309–321.

Ray, K. 2002. Image and Reality: Indian Diaspora Women, Colonial and Post-colonial Discourse on Empowerment and Victimology. In *Women and the Colonial Gaze*, edited by T. Hunt and M. Lessard, 135–147. New York: New York University Press.

Rich, A. 1980. Compulsory Heterosexuality and Lesbian Existence. *Signs: Journal of Women in Culture and Society* 5(4): 631–660.

Ritzer, G. 1992. *Contemporary Sociological Theory*. New York: McGraw-Hill.

Robertson, G. 1896. *The Kafirs of the Hindu Kush*. London: Lawrence and Bullen.

Robinson, J. 1994. White Women Researching/Representing "Others": From Antiapartheid to Postcolonialism? In *Writing Women and Space*, edited by A. Blunt and G. Rose, 197–226. New York: The Guilford Press.

Rose, N. 1999. *Powers of Freedom: Reframing Political Thought*. Cambridge: Cambridge University Press.

Ruddick, S. 1996. Constructing Difference in Public Space: Race, Class and Gender as Interlocking Systems. *Urban Geography* 17(2): 132–151.

Sahgal, G. and N. Yuval-Davis. 1992. Fundamentalism, Multiculturalism, and Women in Britain. In *Refusing Holy Orders: Women and Fundamentalism in Britain*, edited by G. Sahgal and N. Yuval-David, 1–25. London: Virago Press Ltd.

Said, E. 1978. *Orientalism: Western Representations of the Orient*. New York: Vintage Books.

————. 1981. *Covering Islam: How the Media and the Experts Determine How We See the Rest of the World*. New York: Pantheon Books.

————. 1989. Representing the Colonized: Anthropology's Interlocutors. *Critical Inquiry* 15(2): 205–225.

————. 1993. *Culture and Imperialism*. London: Chatto and Windus.

Sarup, M. 1994. Home and Identity. In *Travelers' Tales: Narratives of Home and Displacement*, edited by G. Robertson, M. Mash, L. Tickner, J. Bird, B. Curtis, and T. Putnam, 93–104. London: Routledge.

Saunders, R. 1999. Gender, Colonialism, and Exile: Flora Annie Steel and Sara Jeannette Duncan in India. In *Women's Writing in Exile*, edited by M. Broe and A. Ingram, 304–324. Chapel Hill: The University of North Carolina Press.

Schevill, M., J. Berlo, and E. Dwyer. 1991. *Textile Traditions of Mesoamerica and the Andes: An Anthology*. New York: Garland.

Schneider, J. 1980. Trousseau as Treasure: Some Contradictions of Late Nineteenth-Century Change in Sicily. In *Beyond the Myths of Culture: Essays in Cultural Materialism*, edited by E. B. Ross, 323–358. New York: Academic Press.

———. 1987. The Anthropology of Cloth. *Annual Review of Anthropology* 16: 409–448.

Schomberg, R. 1935. *Between the Oxus and the Indus*. London: Martin Hopkinson.

Schor, J. and F. Schor. 1952. At World's End in Hunza. *National Geographic Magazine* 104(4): 485–518.

Scott, James. 1986. Everyday Forms of Peasant Resistance. In *Everyday Forms of Peasant Resistance in South-East Asia*, edited by J. Scott and B. Tria Kerkvliet, 22–31. London: Frank Crass.

———. 1990. *Domination and the Arts of Resistance: Hidden Transcripts*. New Haven: Yale University Press.

Scott, Joan. 1992. Experience. In *Feminists Theorize the Political*, edited by J. Butler and J. Scott, 22–40. New York: Routledge.

Scott, P. 1998 [1966]. *The Raj Quartet*. Chicago: University of Chicago Press.

Scully, P. 1995. Rape, Race and Colonial Culture: The Sexual Politics of Identity in the Nineteenth Century Cape Colony, South Africa. *American Historical Review* C 2: 335–359.

Sharpe, J. 1991. The Unspeakable Limits of Rape: Colonial Violence and Counter-Insurgency. *Genders* 10: 25–46.

———. 1993. *Allegories of Empire: The Figure of Woman in the Colonial Text*. Minneapolis: University of Minnesota Press.

Sharpley-Whiting, T. D. 1999. *Black Venus: Sexualized Savages, Primal Fears, and Primitive Narratives in French*. Durham: Duke University Press.

Shaw, I. 1990. *Pakistan Handbook*. California: Moon Publications Inc.

Shaw, I. and B. Shaw. 1993. *Pakistan Trekking Guide: Himalayas, Karakorum and Hindukush*. Hong Kong: Vanguard Books Ltd.

Shipton, E. 1938. *Blank on the Map*. London: Hodder and Stoughton.

Shohat, E. 1991. Gender and Culture of Empire: Toward a Feminist Ethnography of the Cinema. *Quarterly Review of Film and Video* 13(1–3): 45–84.

Shohat, E. and E. Stam. 1994. *Unthinking Eurocentrism: Multiculturalism and the Media*. Minneapolis: University of Minnesota Press.

Sibley, D. 1995. *Geographies of Exclusion: Society and Difference in the West*. London: Routledge.

Sidhwa, B. 1993. *An American Brat*. New Delhi: Penguin Books.

Sinha, M. 1992. "Chathams, Pitts, and Gladstones in Petticoats": The Politics of Gender and Race in the Ilbert Bill Controversy, 1883–1884. In *Western Women and Imperialism: Complicity and Resistance*, edited by N. Chaudhuri and M. Strobel, 98–116. Indianapolis: Indiana University Press.

Smith, D. 1988. *The Everyday World as Problematic: A Feminist Sociology*. Milton Keynes: Open University Press.

Smith, S. 1987. Fear of Crime: Beyond a Geography of Deviance. *Progress in Human Geography* 11: 1–23.

Smyth, E., S. Archer, P. Bourne, and A. Prentice. 1999. *Challenging Professions: Historical and Contemporary Perspectives on Women's Professional Work*. Toronto: University of Toronto Press.

Spear, P. 1963. *The Nabobs*. London: Oxford University Press.

Spivak, G. 1985. Three Women's Texts and a Critique of Imperialism. *Critical Inquiry* 12(1): 243–261.

———. 1988. Can the Subaltern Speak? In *Marxism and the Interpretation of Culture*, edited by C. Nelson and L. Grossberg, 271–313. London: Macmillan.

Sprinker, M. 1992. *Edward Said: A Critical Reader*. Cambridge: Blackwell Publishers.

Spurr, D. 1993. *The Rhetoric of Empire: Colonial Discourse in Journalism, Travel Writing, and Imperial Administration*. Durham: Duke University Press.

Stacey, J. 1991. Can There Be a Feminist Ethnography? *Women's Studies International Forum* (11): 21–27.

Stanley, L. and S. Wise. 1983. *Breaking Out: Feminist Consciousness and Feminist Research*. London: Routledge.

Steel, F. A. and G. Gardiner. 1905. *The Complete Indian Housekeeper and Cook*. London: Heinemann.

Steinmetz, G. 2002. Precoloniality and Colonial Subjectivity: Ethnographic Discourse and Native Policy in German Overseas Imperialism, 1780s–1914. *Political Power and Social Theory* 15: 135–228.

———. 2003. The Devil's Handwriting: Precolonial Discourse, Ethnographic Acuity and Cross-Identification in German Colonialism. *Comparative Studies in Society and History* 45: 41–95.

Stoler, A. 1989. Making Empire Respectable: The Politics of Race and Sexual Morality in Twentieth-Century Colonial Cultures. *American Ethnologist* 16: 634–660.

———. 1995. *Race and the Education of Desire: Foucault's History of Sexuality and the Colonial Order of Things*. Durham: Duke University Press.

———. 1997. Carnal Knowledge and Imperial Power: Gender, Race and Morality in Colonial Asia. In *The Gender/Sexuality Reader: Culture, History and Political Economy*, edited by R. Lancaster and M. Di Leonardo, 13–36. New York: Routledge.

Stone-Mediatore, S. 1998. Chandra Mohanty and the Revaluing of 'Experience.' *Hypatia* 13(2): 116–133.

Streefland, P., S. Khan, and O. Van Lieshout. 1995. *A Contextual Study of the Northern Areas and Chitral*. Gilgit: Aga Khan Rural Support Program.

Strobel, M. 1991. *European Women and the Second British Empire*. Indianapolis: Indiana University Press.

Sturma, M. 2002. *South Sea Maidens: Western Fantasy and Sexual Politics in the South Pacific*. Westport: Greenwood Press.

Sutcliffe, B. 1999. The Place of Development in Theories of Imperialism and Globalization. In *Critical Development Theory*, edited by R. Munck and D. O'Hearn, 135–154. London: Zed Books.

Swartz. D. 1997. *Power and Culture: The Sociology of Pierre Bourdieu.* Chicago: University of Chicago Press.

Tarlo, E. 1996. *Clothing Matters: Dress and Identity in India.* Chicago: University of Chicago Press.

Tarr, Y. 1972. *Bread and Soup Cookbook.* New York: The New York Times Book Co. Inc.

Tessler, M. and J. Jesse. 1996. Gender and Support for Islamist Movements: Evidence from Egypt, Kuwait, and Palestine. *The Muslim World* 86(2): 200–217.

Thompson, J. 1984. *Theory of Ideology.* Cambridge: Cambridge University Press.

Tönnies, Ferdinand. 1957 [1887]. *Community and Society: Gemeinschaft und Gesellschaft.* Trans. Charles P. Loomis. Ann Arbor: The Michigan State University Press.

Tranberg Hansen, K. 1989. *Distant Companions: Servants and Employers in Zambia, 1900–1985.* Ithaca: Cornell University Press.

———. 1992. White Women in a Changing World: Employment, Voluntary Work, and Sex in Post–World War II Northern Rhodesia. In *Western Women and Imperialism: Complicity and Resistance*, edited by N. Chaudhuri and M. Strobel, 247–268. Indianapolis: Indiana University Press.

Tyler, S. 1986. Post-modern Ethnography: From Document of the Occult to Occult Document. In *Writing Culture: The Poetics and Politics of Ethnography*, edited by J. Clifford and G. Marcus, 122–140. Berkeley: University of California Press.

Valentine, G. 1989. The Geography of Women's Fear. *Area* 21(4): 385–390.

———. 1991. Women's Fear and the Design of Public Space. *Built Environment* 18: 288–303.

Van den Abeele, G. 1992. *Travel as Metaphor: From Montaigne to Rousseau.* Minneapolis: University of Minnesota Press.

Van Maanen, J. 1990. Escape from Modernity: On the Ethnography of Repair to the Repair of Ethnography. *Human Studies* 13(3): 275–284.

Vigne, G. 1987 [1844]. *Travels in Kashmir, Ladakh, Iskardo, the Countries Adjoining the Mountain Course of the Indus, and the Himalaya, North of the Punjab.* Karachi: Indus Publications.

Visser-Hooft, J. 1926. *Among the Kara-Korum Glaciers in 1925.* London: Edward Arnold.

Visweswaran, K. 1994. *Fictions of Feminist Ethnography.* Minneapolis: University of Minnesota Press.

Ware, V. 1992. *Beyond the Pale: White Women, Racism, and History.* Verso: London.

———. 1996. Defining Forces: "Race," Gender and Memories of Empire. In *The Post-colonial Question: Common Skies, Divided Horizons*, edited by I. Chambers and L. Curti, 142–156. London: Routledge.

Weber, C. 2001. Unveiling Scheherazade: Feminist Orientalism in the International Alliance of Women, 1911–1950. *Feminist Studies* 27(1): 125–157.

Weedon, C. 1987. *Feminist Practice and Poststructual Theory*. Oxford: Blackwell Publishers.

Weiner, A. and J. Schneider. 1989. *Cloth and the Human Experience*. Washington, DC: Smithsonian Institution Press.

Weyland, P. 1997. Gendered Lives in Global Spaces. In *Space, Culture and Power: New Identities in Globalizing Cities*, edited by A. Oncu and P. Weyland, 82–97. London: Zed Books.

Wiegman, R. 1993. The Anatomy of Lynching. In *American Sexual Politics: Sex, Gender and Race Since the Civil War*, edited by J. Fout and M. Shaw Tantillo, 233–245. Chicago: Chicago University Press.

Williams, R. 1988. *Keywords: A Vocabulary of Culture and Society*. London: Fontana.

Wilson, E. 1987. *Adorned in Dreams: Fashion and Modernity*. London: Virago.

Wilson, K. 2004. Empire, Gender and Modernity in the Eighteenth-Century. In *Gender and Empire*, edited by P. Levine, 14–45. Oxford: Oxford University Press.

Wilton, R. 1998. The Constitution of Difference: Space and Psyche in Landscapes of Exclusion. *Geoforum* 29(2): 173–185.

Wolff, J. 1993. On the Road Again: Metaphors of Travel in Cultural Criticism. *Cultural Studies* 7(1): 224–239.

Woodman, D. 1969. *Himalayan Frontiers: A Political Review of British, Chinese, Indian and Russian Rivalries*. New York: Federick A. Praeger.

Workman, F. 1901. Amid the Snows of Baltistan. *Scottish Geographical Magazine* 17: 74–86.

World Tourism Organization. 2001. Tourism Highlights 2001. Internet edition, http://world-tourism.org. Accessed June 20, 2002.

Wright, M. 1988. *Under Malabar Hill: Letters from India, 1928–1933*. London: British Association for Cemeteries in South Asia.

Yazbeck Haddad, Y. and J. Smith. 1996. Women in Islam: "The Mother of all Battles." In *Arab Women: Between Defiance and Restraint*, edited by S. Sabbagh, 137–150. New York: Olive Branch Press.

Yeganeh, N. 1993. Women, Nationalism and Islam in Contemporary Political Discourse in Iran. *Feminist Review* 44: 3–18.

Yeoh, B. and L. Khoo. 1998. Home, Work, and Community: Skilled International Migration and Expatriate Women in Singapore. *International Migration* 36(2): 159–186.

Yeoh, B., L. Khoo, S. Huang, and K. Willis. 2000. Global Cities, Transnational Flows and Gender Dimensions: The View from Singapore. *Tijdschrift voor Economische en Sociale Geografie* 91(2): 147–158.

Young, I. M. 1990. The Ideal of Community and the Politics of Difference. In *Feminism/Postmodernism*, edited by L. Nicholson, 300–323. New York: Routledge.

Young, R. 1990. *White Mythologies: Writing History and the West.* London: Routledge.

———. 1995. *Colonial Desire: Hybridity in Theory, Culture and Race.* London: Routledge.

———. 2001. *Postcolonialism: An Historical Introduction.* Oxford: Blackwell Publishers.

Younghusband, F. 1892. Journeys in the Pamirs and Adjacent Countries. *Proceedings of the Royal Geographical Society* 14(4): 205–234.

Yuval-Davis, N. 1997. *Gender and Nation.* London: Sage.

Yuval-Davis, N. and F. Anthias. 1989. *Woman-Nation-State.* New York: St. Martin's Press.

Zoller Booth, M. 2002. Education for Liberation or Domestication? Female Education in Colonial Swaziland. In *Women and the Colonial Gaze*, edited by T. Hunt and M. Lessard, 174–187. New York: New York University Press.

Index